TRIBES, FOREST AND SOCIAL FORMATION IN INDIAN HISTORY

Tribes, Forest and Social Formation in Indian History

Edited by

B.B. CHAUDHURI
ARUN BANDOPADHYAY

Indian History Congress

MANOHAR
2026

First published 2004
Reprinted 2026

ISBN 978-81-7304-551-6

Published by

Ajay Kumar Jain for
Manohar Publishers & Distributors
4753/23 Ansari Road, Daryaganj
New Delhi 110 002

Printed and bound in India

Contents

Preface

Since the Patiala Session (1999), the Indian History Congress (IHC) has been organizing Panels to discuss particular themes for which scholars are specially invited. This has been possible because of the financial help from various institutions. My predecessor Professor Shireen Moosvi was successful in publishing the proceedings of six such Panels, which were very well received. At the Bhopal Session (2001) one Panel was organized on 'Tribes, Forest and Social Formation in Indian History', sponsored by the Department of Forest, Government of Madhya Pradesh. Scholars working in the related field were invited to present papers. This volume is a compilation of the papers presented in the Panel and a few papers relevant to the theme published in the earlier proceedings of the Indian History Congress.

Special thanks are due to Professors B.B. Chaudhuri and Arun Bandopadhyay who jointly undertook the burden to edit the volume. The volume obviously does not claim a comprehensive coverage of the theme as indicated by its title. But the introduction of the editors and the contributions help us in broadening our understanding about the transformation of the tribal society in the context of the environmental and other social issues.

We are obliged to many for the help received in organizing the Panel and for being able to publish the present volume. Mr L.K. Joshi, former Vice-Chancellor, M.P. Bhoj (Open) University and Mr. Sudip Banerjee, IAS, Government of Madhya Pradesh, helped us enormously to raise funds for making the Panel possible. Professors Irfan Habib and Shireen Moosvi were constant source of inspiration and support both for organizing the Panel as well as for the publication of the volume. We are obliged to Professor Suvira Jaiswal for kindly presiding over the session. We are thankful to all the contributors to the volume, the participants in the Panel discussion and especially the editors for their diligence. We are specially obliged to Mr Ajay Kumar Jain of Manohar for publishing the volume in a short time.

Secretary, Indian History Congress RAMAKRISHNA CHATTERJEE
1 December 2003

Introduction

B.B. Chaudhuri and Arun Bandopadhyay

> The tribal situation in India cannot be viewed in isolation from the wider contexts of the themes and strategies operating in national life. . . . The emphasis on the necessity of relating tribal problems to national contexts and on linking trends in tribal India to trend pervading the entire society does not even remotely imply that there are no problems which are uniquely tribal.
>
> S.C. DUBE (1972)[1]

> Contemporary critics of ecological romanticism have re-looked at the historical constructions of tribal traditions and customs in earlier works. However, few scholars discuss the implications of this fact for the nature of the community itself. In other words, if community was dependent on the production relations and networks for survival then it would also be necessary to see the manner in which different groups within the community were integrated within the larger political economy. . . . (The critics) challenge some existing pre-conceived notions about tribal identity, what it represents and its relationship with land and forest use pattern.
>
> ARCHANA PRASAD (2003)[2]

There have been significant changes in recent years in the research and emphasis on the historical relationship between tribes and forests in India and the variations of social formation emanating from that. Though this relationship may be meaningfully viewed from long historical perspectives over centuries, the colonial official perception on the subject has become an oft-cited, referring point. However, it has been argued that the perception underwent modifications over time: the view of the tribal world changed from the early colonial administrative idea as 'an adjunct to the counter-insurgency measures of the state' to an anthropological, 'improved understanding of tribal society' in the twentieth century.[3] In other words, there were destabilizing effects of British rule on a relatively unchanging tribal life, consequent upon colonial forest regulations curtailing their customary rights enjoyed since the pre-colonial era. The political and social stability of the tribal society in India appear to be the core issues of the discussion.

Strangely enough, some of the issues remained unchanged even after independence. When, for example, a major seminar on the 'Tribal Situation in India' by the Indian Institute of Advanced Study, Shimla was organized as late as July 1969, with participants including anthropologists, sociologists, historians, social workers, tribal leaders and administrators, Professor M.N. Srinivas as Chairman set the tone of the discussion: '. . . anthropological and sociological research can no longer be regarded as something peripheral or irrelevant to the crucial pursuit of the twin objectives of economic development and political stability'.[4] Professor Niharranjan Ray elaborated these ideas from the point of view of Indian nationalism.[5] He made five important observations, each of which being symptomatic of the prevalent thinking on the subject. First, the communities of peoples generally grouped as 'tribes' are nothing but *janas* or peoples, just like peoples of other territorial and cultural regions of India. Second, there is a long tradition of integration of these communities within Indian history and culture, gradually drawing them into the fold of the *jati* system and placing them at different hierarchical levels in it. Thirdly, the British colonial imperial policy of war, conquest, administration and legislation exposed these communities to the much quicker tempo of modernization and to a new legal and economic system than they could accommodate. These definitely gave birth to stresses and tensions amongst them. Fourthly, independent India had in it the implantation of the divisive seeds of 'criminal tribes', which were replaced by the phrase 'denotified tribes' but it hardly changed the situation. Finally, independence, democracy and adult franchise have released many submerged social forces, which were likely to play an important role in their active life. The old order was fast deteriorating.[6]

A significant shift in this analytical framework started in the 1980s, though it became more pronounced in the 1990s. An instrument of this change was the emergence of environmental history as a sub-discipline of study. The other change was noticeable in the renewed emphasis on the study of communities including tribal communities with reference to their relative degree of isolation, autonomy and linkages with outer world. *Subaltern Studies* and their critiques have been useful in comprehending the nature of these twin areas of change.

Partly as a result of the ongoing debate on the state of environment in India in the1980s but also partly as a follow-up of the historical debate on the nature of peasant and tribal movements in India, ecological resistance came to the fore of historical discussion.[7] As early as 1986, K.S. Singh argued that 'tribal movements were both agrarian and forest based' because of their crucial dependence on both forests and land.[8] A few published volumes of *Subaltern Studies* had already included articles on forest-based tribal politics and movements,[9] and historians like B.B. Chaudhuri became more interested in studying the nature of the millenarian movement of the tribals.[10] It was, however, the publication of Ramachandra Guha's *The Unquiet Woods* in 1988 that made the major difference in the historical studies of ecological change and resistance in India.[11]

A corpus of literature quickly developed in the 1990s on the history and management of forests and forest communities in their multi-farious dimensions. A good number of questions generally ignored earlier were now raised but there was no uniformity in the answers. Altogether different positions were often taken by the researchers working in the field. The questions also multiplied over time. To what extent were the tribes or communities residing in forests living in isolation of the peasant society? How far were they prudent in the use of forest resources in the pre-colonial era? How did state formations affect natural resource exploitation and the respective social formations of communities over time? An extension of the same question: What was the exact nature of the colonial watershed in ecological terms? How and to what extent have the issues related to forest management and community rights been tackled in post-colonial era from ecological, technological and participatory demo-cratic perspectives?

Works published in the wake of Guha's volume (1988) include writers like Vandana Shiva, V.F. Haimendorf, Bina Agarwal, Richard Grove, David Arnold, Mahesh Rangarajan, Nandini Sunder, Sumit Guha, K. Sivaramakrishnan, Ajay Skaria and Arun Agarwal.[12] Ramachandra Guha and Madhav Gadgil subsequently developed their pioneering view of the ecological history of India.[13] One effect of contributions of these authors is that a general rift has developed between the concerns of agrarian history and those of environmental history, so much so that the latter's growth has been accompanied in the decline of interest in the former in many ways.[14] The other effect is that the tribal history of India is no longer considered to highlight merely themes of economic development and national integration but to dwell on a number of complex issues like identity, class and gender, ecology and equity, state formation and ethnic transformation, and participatory management of forest resources by the community.

The writers are, however, divided on a number of key issues. One such issue is the paradigm of nature and culture as the valid framework of analysis in environmental history. Though many writers like Gadgil, Guha and Shiva broadly subscribe to this kind of analysis, scholars like Sivaramakrishna differ. He has made two criticisms: (a) nature-culture dualism is to be simplified, and the concept of power should be included in it so much so that state-making may be made an additional, useful factor of analysis and (b) pre-occupation with colonial watershed is far-fetched. Sivaramakrishna examines colonial forestry in three different ways: (1) as a set of material technologies imposed on trees, grasses and wild animals (2) as a legal regime aimed at appropriation and monopoly in the extraction of natural rents and (3) as a system of rational knowledge (scientific forestry) that became the site of a struggle amongst technocrats and bureaucrats. However, because of the limitations of the colonial statemaking, Sivaramakrishna continues, forestry as a technology could neither consolidate its hold on the most valued tree species nor was able to codify its rapid regeneration. Secondly, forestry as a legal regime never achieved its consistency in application—exclusive claims were forsaken and regulations were relaxed. Finally, 'forestry as a system of professional knowledge failed to garner respect and authority because many colonial elites remained unconvinced of its scientific pedigree'.[15]

Another recent controversy relates to the nature of ecological romanticism as evident from certain studies in tribal culture. It has been argued that ecological romanticism, originally rooted in English opposition to Industrial Revolution in the nineteenth century, made a general defence of tribal customs and traditions in ecologic terms and often took them as alternatives to the ethos of modern society. 'More importantly, it also formed the basis of an environmental ideology that became dominant amongst Gandhian movements of the twentieth century'.[16] An extension of the same argument is found in the assumption that tribal practices, as reflected in their cultural and social life, were ecologically friendly and that their ethnic identities mirrored their cultural distinctiveness. The biographer of Verrier Elwin even went to the extent of saying that 'the Gond, the Baiga and the Agaria all believed that the past had been better than the present, that there had truly once been a Golden Age when their kings ruled. . .'.[17] Undoubtedly, there is an argument against ecological romanticism, but the problem is that a valid space for environmental history and environmental movements is also to be found out.

Considering the importance of the subject and the state of historical debate within it, the Indian History Congress organized a Panel on 'Tribes, Forest and Social Formation in Indian History' at its Bhopal Session in 2001. However, the publication could not be made immediately as the contributors took time to revise their papers. Unfortunately, some of them did not reach us at all, despite our repeated reminders. Three articles pertaining to the subject have been reproduced from the Proceedings of the Indian History Congress in recent years.[18] The volume contains twelve papers.

II

B.D. Chattopadhyaya raises certain interesting questions on the very conceptualization of forest, forest-dwellers and forest kingdoms in early India in terms of differences between simple and complex societies. He argues that the proximity between forests and human communities, both physical and attitudinal, sharply changes as such communities move towards greater complexity, best exemplified by the emergence of the state. To the state, the 'forest' is a necessary space, but not necessary in the same way in which it is to those who are forest-dwellers. The state's perception of the forest and those who dwell in it is defined by those who formulate and uphold social codes in consonance with organization of complex societies. For instance, Chattopadhyaya mentions Rama's sermons to dying Bali in the *Ramayana* as an example of the attitude of one who represented the political order of a complex society. At the material level, the state would draw on the resources available in the forests and derive advantage, by various means from its domination over the forest dwellers. However, he argues, the expansion of the frontiers of complex societies could bring about rupture, as a historical process, in the social organization of the forest-dwellers.

While recapitulating his thinking on forests, forest-dwellers and ecological history, K.S. Singh further clarifies some of the concepts in the process. For him, the meaning of the word jungle varied from the ancient view of it as mysterious and the poetic perception as 'dark and deep' to the tribal people's own view of reality. Historically speaking, this world represents at the same time continuity and dichotomy, conflicts and integration, and male-female division of work as usual. There was no marked dichotomy between forest and agriculture in the early period, but the tribes gradually developed an extraordinary knowledge of environment, of plants and crops and a sensitivity to resources like soil and water. However, they were at the same time the greatest reclaimers of land, doing all the hard work of breaking soil, clearing forests and removing vegetation. There is no linear process of the transformation of tribes into peasants as assimilation was never complete. However, there has been a gradual, almost unexplained surrender of women's role in hunting to men under the influence of patriarchal norms over tribal society. Finally, he is an ardent believer of colonial watershed as the 'forest became for the first time a resource controlled and regulated for the benefit of the British empire and British capital'. Forest-based agitations in India were a consequence, which reached its most articulate form in the post-colonial era.

Kesavan Veluthat questions the very process of transformation from tribe to caste in the long historical perspective for the Koragas of South Canara. The transformation has been traced from well before the colonial period with reference to the many narratives of two contending accounts: Koraga legends of Hubbashika (the Koraga king of olden time) and the Brahmanical stories of Koraga subjugation. The differences are obvious, but they are suggestive of a relationship of the Koragas with other castes from the pre-colonial period in such a way that a pre-caste existence seems impossible. Indeed, the Koragas, officially labelled as a 'primitive tribe', are actually living as a caste with all its features in a caste society.

Rajan Gurukkal analyses physiographical assemblage of five landscape ecotypes in pre-Pallavan south India (from the first millennium BC to about the fifth century AD) with special reference to forest and the interactive social co-existence of various tribes with two or more extant economies. As most landscape ecotypes (with the possible exception of the terrain of wet-rice agriculture) were within forest eco-system, and the three unevenly evolved chiefdoms (the *Velir*, *Kilar* and *Ventar*) were connected with forest and forest lands, the gradual dissolution of social formation was noticeable in the expansion of wet-rice cultivation, slow disintegration of tribes into domestic segments of hereditary arts and crafts, and the formation of agrarian villages at the end of the period.

S.K. Panda focuses on the cross-cultural transmission between the tribals and the Hindus in Orissa mainly during the early medieval period with special reference to the development of the cult of Jagannatha and the Shakti worship. Brahmanization and acculturation of the tribals, and the cross-cultural assimilation of their culture in Vishnuite Hinduism went hand in hand. A large number of land grants, given to the Brahmans, finally led to the extension of agriculture in forest areas.

B.B. Chaudhuri distinguishes between the two terms—*Adivasis* and *Aranyakas*—respectively as tribals and forest-dwellers, without taking an essentialist position but maintains that their formation was historically determined. In characterizing the polity and economy of the respective groups in pre-colonial and colonial India, Chaudhuri analyses the nature of forest control and forest use by pre-colonial rural communities and governments, along with the nature of pre-colonial tribal village's interconnections with the wider polity and economy. He does not find the arguments based on cultural tradition and communal control very useful in explaining the exclusion of a specific village space from cultivation for ever, and thinks that the role of Mughal government (and virtually all pre-colonial governments) in conservation was limited and at best indirect. Similarly, tribal village's interconnection with the external world was a long-drawn-out process, so much so that state-formation became the most enduring and decisive form of the *adivasi* village's integration with the larger polity even before the colonial era. For him, the ecological implications of colonial watershed are to be found mainly in degrees of use and interference. In the pre-colonial era, the use of forest resources by the community and the state remained limited on the one hand, and there was virtual non-interference in the tribal village organization on the other. The changes in the twin areas are phenomenal in the colonial era.

Archana Prasad also raises certain serious questions about the nature of tribal livelihood systems in late pre-colonial and colonial era and focuses on its implications in the post-colonial period. Instead of seeing the exclusion of local rights and the destruction of the local livelihood systems as the dramatic result of the modern scientific forestry project, she delineates the wider process of tribal destitution through the spread of industrial capitalism in Chhattisgarh (for the Baiga and the Gonds). She argues that the Gonds in the pre-colonial period survived by living in the jungle and doing semi-permanent cultivation with the plough. It was only at the turn of the eighteenth century (post-Raghujee period) that their dependence on forests (and, therefore, on gathering) increased considerably. She further argues that the Baiga subsistence in the colonial period depended on a combination of gathering, hunting and *bewar* (a Baigani term for shifting cultivation), but the linkages amongst these three activities became precarious with colonial forest policy in the nineteenth century. It is because of this connection between *bewar* and other practices that a minimum level of forest diversity had to be maintained. However, different aspects and facets of colonial agrarian policy just did the opposite. With the complete destruction of the productive capacity of the tribal people and their conversion into cheap providers of labour and raw materials in a capitalist world increasingly linked with global markets, tribal identity politics became visible since the twentieth century. However, the irony of the matter is that an aggressive tribal identity got strengthened even as the material basis of its existence started to disappear and pre-capitalist forms gradually got incorporated in the colonial and capitalist mode of production.

Amar Farooqui focuses on the particular case of the transformation of the Pindaris—a formidable tribe of the Malwa region—against the backdrop of a continuous historical process of social differentiation in the eighteenth and nineteenth centuries. This process was linked to state formation, empire building and the search for military recruits. Instead of identifying the Pindaris as completely isolated, Farooqui talks of 'some exchanges between them and settled agriculturists' in the Mughal period. But their real integration began when many of them found a place in the Maratha military organization. Prior to the British conquest of the region in 1818, the Pindari chiefs retained holdings in the Narmada Valley, and two of them—namely, Chitu and Karim Khan—became very prominent, significantly receiving land grants from Scindia. What is more notable, the Pindari leaders retained outposts in the Narshimapur-Hoshangabad-Dewas stretch of the Narmada Valley, not merely to have a safe refuge but mainly to enlist followers from among the tribal communities (the Gonds and the Bhils), some of whom had already been transformed into tribal bands of plunderers in the central Indian forests. After 1805, the Pindari participation in Maratha campaigns assumes new significance, as both Yashwant Rao Holkar and Daulat Rao Scindia forged an alliance with them, leading to their acquiring a limited share in political power. The suppression of the Pindaris put an end to their state formation, and thousands of unemployed Pindaris with some military training became the central Indian banditry after 1818.

K.T. Thomas brings us to south India when he tries to evaluate the official policy towards forests and conservation in the native state of Cochin in the nineteenth and twentieth centuries. He does not find any great difference in the policy in terms of exploitation and differentiation, compared to the colonial policy in British India. However, colonialism as a whole had a role in the deforestation of Cochin. Finally, the limited policy of reforestation and conservancy was not influenced by ecological considerations but by economic and commercial motives.

P.K. Shukla reaches similar conclusions in his study of Dubia Gossain movement in the tribal region of Chotanagpur under the Raj in the 1870s and 1880s. Coming in the wake of Kherwar movement, Dubia Gossain movement was mainly directed against the colonization of the area. It was an agrarian issue, and not a religious one. The demands of the Santals included restoration of their lost lands, stoppage of the enhancement of rent, and an end of the encroachment to their grazing land and forest wealth. However, they made use of their religious myths and legends to strengthen struggle.

Atlury Murali makes an overall study of forests, development ideology and peasant-tribal life in the Madras Presidency over the last two hundred years. Beginning with the strong historiographical tradition in the studies of local society including peasant and tribal world, Murali argues that livelihood factor was not a limiting element of local economy or village community. He thinks that the language of exclusion in forest rights was not so pronounced as the colonial regime's concern for timber and teak was in the early phase, but it was definitely so in the second half of the nineteenth century, when modernization, commercialization, and the widening of the market were associated with it. He, therefore, highlights the relationship between forests, the colonial paradigm of conservation as dictated by the Western notion of development and the rights of the natives as characterized by the peasant-tribal life world in south India.

Arun Bandopadhyay takes up some of these issues in post-colonial era and discusses them with reference to the management of Common Property Resources (CPRs) in Tamilnadu and West Bengal. Bandopadhyay contends that the management and productivity of CPRs in post-colonial India are associated with a variety of issues like environment and ecology, poverty and equity, rural development, bureaucratic perception and participatory democracy. The three perspectives—ecological, technological, participatory—may work as useful tools of analysis, their significance being in the complex interplay, overlapping and even interdependence.

III

The essays presented in this volume reflect to a large extent the existing state of research in a rapidly changing area of historical interest. They are not definitive, and they do represent multiple viewpoints of a lively controversy. We have noted that significant shifts have occurred in the scholarly discussion on the subject in the last thirty years; old issues of economic development and social integration have given place to the new ones of community, culture, power, class and identity in recent times. All history is dated, and so is the scholarly pre-occupation of today. But what is more important, the old issues are not completely dead; they may be resurrected with new meanings and forms at any time. The state of research does not enable us to give a final judgement on any of the issues. Take, for example, the question of power as a possible critique of the construction of nature and culture in environmental history. David Hardiman in a recent study has argued that shifts of power in tribal society from the locally dominant groups to others were important for the forest society from the pre-colonial period down to the end of the colonial era, transforming the forest in the process as well as turning the erstwhile ruling community, the Bhils, into a subaltern community.[19] However, Hardiman himself has admitted at the end of his pre-colonial section that the Bhils 'had a strong affinity with these woods and hills—their home as well as their place of refuge—and any destruction which they carried out was on so small a scale as to make very little difference to the environment as a whole'.[20] Similar are the conclusions of some of the essays contained in the volume, though they make a major revision of the older view on the pre-colonial formulations on the subject.

Some of our essays also do significantly contribute on the nature of ecological romanticism as found in recent literature. While taking a general position against the essentialist and grossly romantic thinking on the subject, they also search for a valid space for ecological perspectives on it. It seems that the search will continue for long. There is always a scope for reassessment of certain issues or personalities. The impact of Gandhian thought, for example, requires re-evaluation from environmental perspectives.[21] Probably, we have to be more careful in grossly equating a functionalist eco-logic with an espousal of tradition and customs of the community. Gadgil and Guha's *This Fissured Land* is an example. Though generally accused of a bias in favour of ecological romanticism, the book does not make any specific reference to traditional sources, Sanskrit or vernacular, even in its analysis of pre-modern developments.[22] Our gap of knowledge concerning the issues discussed here for the pre-colonial, colonial and post-colonial era is still enormous. We hope this publication will be a step forward in this regard and will pave the way for further research.

NOTES

1. See the Inaugural Address of Professor S.C. Dube delivered at a seminar on 'The Tribal Situation in India' at the Indian Institute of Advanced Study, Shimla in 1969, now included in K. Suresh Singh (ed.), *The Tribal Situation in India* (Shimla, 1972, revd edn 2002), p. 31.
2. Archana Prasad, *Against Ecological Romanticism: Verrier Elwin and the Making of an Anti-Modern Tribal Identity* (Delhi, 2003), pp. xix-xx.
3. See, amongst many historigraphical studies on the subject, Sanjukta Das Gupta's 'Peasant and Tribal Movements in Colonial Bengal: A Historiographic Overview', in Sekhar Bandyopadhyay (ed.), *Bengal: Rethinking History, Essays in Historiography* (Delhi, 2001), p. 79.
4. See M.N. Srinivas, 'Chairman's Address', in K. Suresh Singh (ed.), op. cit., p. 25.
5. See Niharranjan Ray, 'Introductory Address', in ibid., pp. 3-23.

6. Ibid., pp. 20-1.
7. See *The State of India's Environment: A Citizens' Report* (Centre for Science and Environment, Delhi, 1982) and *The State of India's Environment: A Second Citizens' Report* (CSE, Delhi, 1985) and the many discussions that followed in the pages of *Economic and Political Weekly* (hereafter *EPW*) but particularly Ramachandra Guha's 'Ideological Trends in Indian Environmentalism', in *EPW*, 3 December 1988 and 'Social-Ecological Research in India: A "Status" Report', in *EPW*, 15 February 1997.
8. K.S. Singh, 'Agrarian Dimension of Tribal Movements', in A.R. Desai (ed.), *Agrarian Struggles in India after Independence* (Delhi, 1986), p. 166.
9. See, for example, articles of Swapan Dasgupta ('Adivasi Politics in Midnapur, *c.*1760-1921') and Tanika Sarkar ('Jitu Santal's Movement in Malda,1924-1932'), in Ranajit Guha (ed.), *Subaltern Studies IV: Writing on South Asian History and Society* (Delhi, 1985).
10. Chaudhuri's interest in tribal studies was gradual but it decidedly developed in the 1980s. However, most of his articles on the subject were published in the 1990s. See, for example, B.B. Chaudhuri, 'The Story of Tribal Revolt in the Bengal Presidency: the Religion and Politics of the Oraons, 1900-1926', in Adhir Chakraborty (ed.), *Aspects of Scio-Economic Changes and Political Awakening in Bengal* (Calcutta, 1989); 'Tribal Society in Transition: Eastern India, 1757-1920', in Mushirul Hasan and Narayani Gupta (eds.), *India's Colonial Encounter* (Delhi, 1993) and 'Ideology and Organisation of the Millenarian Protest Movements in the Tribal World of Colonial Eastern India', in J.T.O'Connell (ed.), *Religious Movements and Institutions in Modern India* (Delhi, 1999).
11. See Ramachandra Guha, *The Unquiet Woods: Ecological Change and Peasant Resistance in the Himalaya* (Delhi, 1988).
12. See Vandana Shiva, *Staying Alive: Women, Ecology and Development in India* (London, 1989); V.F. Haimendorf, *Life among Indian Tribes: An Autobiography of an Anthropologist* (Delhi, 1990); Bina Agarwal, *A Field of One's Own: Gender and Land Rights in South Asia* (Cambridge, 1994); Richard Grove, *Green Imperialism: Colonial Expansion, Tropical Island Edens and the Origins of Environmentalism* (Cambridge, 1995); David Arnold and Ramachandra Guha, *Nature, Culture, Imperialism: Essays on the Environmental History of South Asia* (Delhi,1995); Mahesh Rangarajan, *Fencing the Forest: Conservation and Ecological Change in India's Central Provinces, 1860-1919* (Delhi, 1996); Nandini Sunder, *Subaltern and Sovereigns: An Anthropological History of Bastar, 1854-1996* (Delhi, 1997); Sumit Guha, *Environment and Ethnicity in India 1200-1900* (Cambridge, 1999); K. Sivaramakrishnan, *Modern Forests: Statemaking and Environmental Change in Colonial Eastern India* (Delhi, 1999); Ajay Skaria, *Hybrid Histories: Forests and Wilderness in Western India* (Delhi, 1999) and Arun Agarwal, *Greener Pastures: Politics, Markets and Community among a Migrant Pastoral People* (Delhi, 1999).
13. See Madhav Gadgil and Ramachandra Guha, *This Fissured Land: An Ecological History of India* (Delhi, 1992) and *Ecology and Equity: The Use and Abuse of Nature in Contemporary India* (Delhi, 1995).
14. 'If the old agrarian history neglected the forests and pastures, environmental history now has banished the peasant fields and farms from the realm of historical concern.' See Neeladri Bhattacharya, 'Introduction' to the Special Issue on Forests, Fields and Pastures in *Studies in History*, Vol. XIV, no. 2 (July–December 1998).

15. See Sivaramakrishna, op. cit., p. 3. For a discussion of his general views on statemaking and environmental change, see pp. 1-28.
16. See Archana Prasad, op. cit., p. xiv.
17. Ramachandra Guha, *Savaging the Civilised: Verrier Elwin, his Tribals and India* (Delhi, 1999), p. 122. Also quoted in *ibid,* p. xv.
18. The three articles, contributed by S.K. Panda, K.T. Thomas and P.K.Shukla, are respectively published in the *Proceedings of the Indian History Congress* in the 55th (1994), 58th (1997) and 62nd (2001) sessions.
19. David Hardiman, 'Power in the Forest: The Dangs, 1820-1940', in David Arnold and David Hardiman (eds.), *Subaltern Studies VIII: Essays in honour of Ranajit Guha* (Delhi, 1994), pp. 89-147.
20. Ibid., p. 105.
21. Ramachandra Guha's view on Gandhi changed over time. Though he did not find any 'alternative perspective' on development in Gandhian literature as late as 1998 (Guha and Juan Martinez–Allier, *Varieties of Envirnmentalism: Essays North and South,* Delhi, 1998), he found in Gandhi 'the basis for social renewal in free India' in 2000 (Ramachandra Guha, *Environmentalism: A Global History,* Delhi, 2000, p. 24). See also Arun Bandopadhyay, *Gandhi on Environment: Essential Writings with an Introduction* (forthcoming).
22. See David Ludden's criticism of Gadgil-Guha's book in *Studies in History*, Vol. X, no. 1 (January-June 1994). See also Arun Bandopadhyay, 'Towards an Understanding of Environmental History of India', in *The Calcutta Historical Journal,* Vol. XVI, no. 2 (July-December 1994), pp. 153-65.

State's Perception of the 'Forest' and the 'Forest' as State in Early India

B.D. Chattopadhyaya

> According to some the land of the *Āryas* is between the river Ganges and *Yamunā*. In this connection, moreover, the Bhāllavins cite this verse (va.1.14):
>
> > The boundary river in the west and land of rising sun in the east—between these as far as the black antelope roams, so far does vedic splendour extend.
>
> *Dharmasūtra* of *Baudhāyana*, 1.2.12
>
> (Patrick Olivelle, *Dharmasūtras*: The Law Codes of *Āpastamba*, Gautama, *Baudhāyana*, and *Vasiṣṭha*, New York, 1999, p. 134.)

There are many ways in which forests have been perceived both in written sources of early India and modern historiography of analyses of such sources.[1] The most general characterization of the forest is derived from its terminological contrast with other types of space; if 'forest' was *araṇya, vana, aṭavi, jāṅgala* and so on in texts, it stood in contrast to areas which were different, such as *kṣetra, janapada, nāḍu* (in Dravidian), etc., suggestive of spaces with greater human civilizational associations but nevertheless deriving many elements from the 'forest'.[2] In fact, despite the contrast, the 'forest' is seen to be complementary to existence in other spaces, encompassing both material and transcendental dimensions of being. However, just as there are many terms regarding the 'forest', there could be many terms regarding what 'forests' represented. For example, in the early Indian medical tradition, 'the communities of living creatures that share the conditions of life in a given place',[3] *jāṅgala* was not simply a variety of the 'forest' but a space in which the combination of men, animals, birds, trees and climate (arid, dry and temperate) constituted a set different from that in forest spaces in *anupa* or marshy areas. Similarly, in early Tamil tradition, suggested by anthologies of poems of the Sangams, a zone such as the *kuriñci*—the zone of forested mountains—also signified the nature of a community at a particular subsistence stage of existence, its subsistence practices, its habitat, its deity and rituals, and the way the community could have had channels of communication with no or marginal forestspaces and their communities.[4] In early Buddhist perception of different ages through which human existence had to pass, perhaps life in the forest was the ideal one; most significantly, this was a condition of life which appears to have been free not only from regular toils and intra-community conflicts, but also from the ceaseless cycle of existence, which was, without exception, full of sufferings.[5]

The initial point then is that there could be varieties of perception of the forests which, although there were differences between them in available perceptions, all seem to have derived, in essence, from their linkage to human existence. This brief paper, however, is not concerned with the total range of how forests were viewed, possibly differently by different communities and in different regions. The title of the paper is suggested by premises of basic differences which existed between simple and complex societies, the hallmark of a complex society being the presence of the institution of the State in it. The assumption, deriving from these differences, would be that forests being intimately connected with the life and culture of simple communities had an altogether different meaning for such communities than what they had for members and institutions of complex societies. However, as a historical process the forest-scape could never remain constant; its boundaries would change, and particularly with the expansion of frontiers of complex society, the forests themselves could be transformed into nuclei of local complex societies. Exploring these meanings and processes in detail would be beyond the scope of a single paper. What we therefore propose to do is to examine a selection of evidence bearing upon the Brahmanical perception of how the state, of which the king was the central element,[6] was expected to view the forest and also how kingdoms, emerging out of forest landscapes, were expected to be organized, institutionally as well as ideologically.

II

To an extent—and not beyond that—the Brahmanical perception of the forest may perhaps be traced to the *Ṛgvedic* hymn to the *araṇyāni*, in which the hymnist, who is also a frightened traveller lost in the forest, supplicates the spirit of the forest in the following hymn, beautifully translated by Wendy Doniger O'Flaherty.[7]

1. Spirit of the forest, spirit of the forest, who seem to melt away, how is it that you do not ask about a village? Doesn't a kind of fear grasp you?
2. When the Chichika bird takes up the refrain from the droning cricket, the spirit of the forest is like a hunter startling the game with his noisy beaters.
3. The spirit of the forest at evening: You think you see cows grazing; you think you see a house; you think a cart is rumbling.
4. Whoever stays in the forest at evening imagines: Someone is calling his cow; someone else is cutting wood; someone is crying out.
5. The spirit of the forest does not kill—not if no one else approaches. She eats sweet fruit and lies down wherever she pleases.
6. Mother of wild beasts, untilled by a plough but full of food, sweet-smelling of perfume and balm—to her, the spirit of the forest, I offer my praise.

The spirit of the forest may be benign, but even so, to one who comes from a space which is a village, an area tilled by plough, where the call of the cow is a familiar sound, in which a house and even the rumbling cart are parts of one's normal experience to him the 'mother of wild beasts' is also like a 'hunter startling the game'.

The *Ṛgvedic* hymn may be taken to understand the mystique of the forest from the perspective of a villager who strayed into it, but one can only guess whether the sacredness of the forest, its liminality and the sharp contrast between the transcendental and the evil both of which it came to represent in Brahmanical tradition had their origin in the mystique of the forest. The *Āraṇyaka* category of texts of the Vedic corpus and the association of the forest with special knowledge communicable only through close contact[8] further delineate the separatedness between the ordinary world and the forests of hermits, ascetics and of special rituals. The assertion that Veda recitations are forbidden in the city and suitable only in areas where supply of firewood and *darbha* grass was assured underlines this separatedness further. And yet, even for a householder, retreat to the forest was essential, because if he were to follow the *āśrama-dharma*, the final stage of his life was to be spent as a forest hermit.[9] 'Clothes made of materials from the wild are prescribed for him. Thereafter, he should roam about, living on roots, fruits, leaves, and grasses, and finally on what he happens to find lying about. . . . From then on he should use only wild produce to offer fire sacrifices, to sustain himself, to attend to guests, and to clothe himself.'[10]

What was the relevance of the apparent physical separatedness of the forest to the hierarchical spatial arrangement of the complex society, and for its members residing in settlement units such as *ghoṣa*, the *grāma*, the *nigama*, the *pura*?[11] Since life in the complex society had to be regulated, not in keeping with community/kinship norms or conduct, but, at least in principle, according to *varṇāśrama-dharma* in which the forest hermit's lifecycle was integrated, forest had to be viewed as a part of the complex society and its ideal social and moral order. This integrality should explain: (i) why the monarchical state had to view the forest as within its domain, (ii) why moral authority related—thereby necessitating the separation between the temporal, with its firm location in varieties of non-forest settlements—to the space[12] which to common members of a complex society was alien and full of unfamiliar elements. The threat to the moral authority associated with the forest was of the nature of opposition to its ideal social order and the constant practices which needed to continue—in the form of sacrifices and other daily rituals. In the Brahmanical perception of the state linking it up with forest space therefore the obligation of the monarch was to make sure that such opposition was crushed. Hence, the early texts, including the epics, are replete with references to repeated attempts to spoil sacrifices, to the agony of the *ṛṣis*, and to the *leitmotif* of the temporal authority fighting against the forces of evil to ensure the supremacy of the moral authority. The references are too many and do not need to be cited, but two points ought to be mentioned before we pass on to the next section.

The first is that the state's perception of the forest did not—and could not—leave out of consideration its potential as a generator of resources of various kinds. It was not only timber, honey, and such animals as elephants which made it essential for the forest to remain within the domain of the king but other types of resources as well.[13]

The threat held out to the *aṭavi* or the forest dwellers by *Aśoka*, who appears desperate in his attempt to integrate his far-flung empire (Afghanistan to Orissa and the Deccan to Kashmir) through propagation of a homogenizing policy of *dhamma*, becomes understandable if the state's need to make the *aṭavi* a proper part of the empire is appreciated. After all, the forest-dwellers did not only represent a space which yielded varieties of resources and the all-important revenue for the state, which, for example, for the Mauryan state the herders and hunters had to part with.[14] The forest-dwellers, being demographically an important part of any state's population, needed to be made to conform to what the state required. For example, the settlements in the complex society, the *grāma* (village) and the *nagara*, required that the primarily forest-dwellers like the *Caṇḍālas* and the *Niṣādas*, along with other communities, could provide services to the complex society and be a part of it, but only marginally as untouchables.[15] The forest-dwellers, even in the separation of their own community existence, could further be made a part of the security strategy for the king. Note for example what the *Arthaśāstra*, the most well-known text on polity in early India, says in the section on *janapadaniveśa* (settlement of countryside):

> He (the king) should establish *antapāladurgas* (fortresses of *antapālas*, i.e. frontier-guards) in the *antas* (frontiers). These gateways (*dvāras*) of *janapadas* should be established as the seats (*adhiṣṭhita*) of the *antapālas*. Between them (i.e. between the *janapadas*), the (spaces) will be protected by the *Bāgurikas*, the *Śabaras*, the Pulindas, the *Caṇḍālas* and those who roam in the forest (*araṇya-cra*).[16]

Those who roam in the forest and the others in this recommendation are in the space 'in between'; they inhere the spaces which lie between the *janapadas* which were the nuclei of the complex society. Those 'in between' were to be tamed and such spaces had to be made subservient to the domain of the state. We shall close this section by citing two types of texts to further illustrate how state authorities with their particular formulated codes for their subjects tried to reach out to those in the forests who had their own codes. Maurya *Aśoka*, in the third century before the common era, enunciated the policy of *dhamma*, the cardinal principles of which were non-violence and amicable family and sectarian relationships. In his major rock edict 13 *Aśoka*, by then a practising Buddhist, expressed genuine repentance over the atrocities committed by him in the *Kaliṅga* war;[17] elsewhere, his edicts mention how he substituted the royal hunting expeditions (*vihāra-yātrā*) with *dhamma-yātrā*. And yet in his major rock edict 13 itself *Aśoka* holds out the threat to the *aṭavi* or the forest-dwellers that in case his exhortations with them failed, he could be compelled to eliminate them.[18] There obviously was incompatibility of *Aśoka*'s non-violent *dhamma* and the practices of the forest-dwellers. It is not only that *Aśoka* imposed a ban on the holding of *samājas* or community festivals which were not necessarily those of forest-dwellers;[19] as an emperor his edicts went much deeper into forest life. Note, for example, what be promulgated in his pillar edict 5:[20]

> When I had been consecrated for twenty-six years I forbade the killing of the following species of animals, namely, parrots, *maināś*, red-headed ducks(?), *cakravaka*-geese, swans, *nandī-mukhas* . . . , pigeons, bats, ants, tortoises, boneless fish, *vedaveyakas*, *pupuṭas* of the Ganges fish, skate, porcupines, squirrels, deer, lizards, domesticated animals, rhinoceroses, white pigeons, domestic pigeons, and all quadrupeds which are of no utility and are not eaten. . . .

The edict of course continues, and the essence of it is in consonance with what *Aśoka* would have liked us to believe, as it is expressed in his Greeko-Aramaic inscription found in Shar-i-Kuna near Kandahar in south-east Afghanistan: 'And the king abstains from [killing] living beings, and other men and those who [are] huntsmen and fishermen of the king . . . have desisted from hunting'.[21]

One can guess that *Aśoka*'s choices regarding non-slaughter and the seasons and cycles of permitted slaughter could hardly be expected to have corresponded to those of the forest-dwellers.

If non-violent emperor *Aśoka*'s *dhamma* had one type of implication for the forest-dwellers, the Brahmanical *dharma* of a monarch, with its legitimate use of *daṇḍa* or punishment, could mean supercession of one kind of moral order by another, i.e. the subordination of a simple community by the authority and might of the complex society. This is how *Vālin*, the monkey-chief of *Kiṣkindhā*, views the conduct of *Rāma* who kills him:[22]

> I did not know that your judgement was destroyed and that you were a vicious evil-doer hiding under a banner of righteousness, like a well overgrown with grass.
>
> I did not know that you were a wicked person wearing the trappings of virtue, concealed by a disguise of righteousness. . . .
>
> I did no harm either in your kingdom or in your city, nor did I insult you; so why did you kill me, an innocent forest-ranging monkey, living only on fruit and roots, when I had joined battle with someone else and was not fighting against you?

It is not that *Vālin* is shown as completely ignorant of Brahmanical *dharma*; he continues:

Virtuous people cannot wear my skin, my fur and bones are forbidden, and my flesh cannot be eaten by people like you who observe the law.
Only five among the five-clawed creatures can be eaten by brahamans and kshatriyas, Raghava: the hedgehog, the porcupine, the lizard, the rabbit, and fifth, the turtle.

And yet, in the end, in his most undesired, and undeserved, deathbed *Vālin* accepts that *Rāma* is justified in killing him. This is because *Rāma* has 'convinced' him, in *Vālmīki*'s text of the *Rāmāyaṇa*.

This earth with its mountains, woods and forests belongs to the *Ikṣvākus*, as does the right of punishing and rewarding its beasts, birds and men.

The right to define righteousness (*dharma*) therefore belonged to an exiled prince from distant Kosala, the seat of the *Ikṣvākus*, and not to the local monkey-chief of *Kiṣkindhā*.

III

Forests do not remain static. They shrink with various types of encroachment upon them, or, which is less likely, they take over deserted *janapadas* and the structures which are left behind, leaving it to the archaeologists to locate them. To illustrate the first process, it has been shown by us elsewhere that even when an extensive tract of forest land with wild animals is granted to *brāhmaṇas*, the reference point for the donated space would be settled villages nearby.[23] The inevitability of the transformation of forest land into settled villages, over a period of time, should the donees choose to make use of the grant, is obvious. As a historical process what appears more significant is evidence of the transformation of not only the physical terrain of forest-dwelling communities but of the structures of the communities themselves. Such transformation could have happened in many ways, but the remaining part of this paper will take up select evidence bearing on forests transformed as kingdoms and associated changes that follow, in keeping with the Brahmanical model of monarchy.

If the *aṭavi* of emperor *Aśoka*'s edict represents the forest-dwellers of the kind mentioned by the *Arthaśāstra*, then surely before the mid-fourth century A.D. some forest-spaces which had earlier remained largely unaffected by the spread of monarchy had started feeling the impact of this model of polity. Gupta ruler Samudragupta's contemporary *Vyāghrarāja* of *Mahākāntāra*, humbled by him, or all the *āṭavikarājas* (forest kings),[24] reduced to the status of servants by him (*paricārakī-kṛta*), would suggest one facet of the emergence of *rājyas* and *rājās* among the forest-dwellers, as, to an extent, the increasing use of the titles *rājā* and *māhārāja* among single members of contemporary *gaṇasaṃghas* who were eliminated by Samudragupta. In fact, the Gupta model appears to have become an ideal model for local-level monarchies, over a wide stretch of this subcontinent; in some areas, as in the Himalayan belt, it is only the post-Gupta period that started experiencing the emergence of local-level monarchical states for the first time[25], accelerating the pace of historical transformation in the context of these areas for the first time.

Much of this transformation took place right in the heart of forest-dwellers' domains. If the *Rāmāyāṇa* provides evidence of hegemony over the forest-dwelling monkeys claimed by a prince in distant exile, late and post-Gupta inscriptions speak, also in contrast to the spirit of the inscriptions of *Aśoka*, of 'kingdoms of the forest' (*āṭavi-rājya*) in the Baghelkhand region of present-day Madhya Pradesh.[26] In fact, transformation of forest-spaces into spaces as nuclei of *rājyas* (kingdoms) had profound implications for local landscape, community and structure of ideology. In historiography, the major points in this transformation which are highlighted are: economic exploitation through transfer of land rights to those who were outsiders and to institutions, and of the right to take total control of the alienated land and people.[27] This, however, offers only a partial view of the wide range of fundamental mutations which were likely to take place in the forest space and among forest-dwellers when the forest space emerged as the nucleus of a 'forest kingdom' which, however, would be a segment of larger political formation to which it would be either subservient or with which it would contest for relative autonomy. At one level, the symbols which are associated with the emergence of a monarchy are generally: the insistence on a respectable genealogy—a genealogy approved both by the epic and Puranic traditions, with particular preference for the solar lineage (*Sūryavaṃśa*); the urge felt for the presence of *brāhmaṇas* in the kingdom; patronage for the construction of temples which enshrined Puranic duties such as *Viṣṇu* (or his incarnation); *Śiva*, *Śakti* or of local deities assuming the identity of a Puranic deity; the assumption of and pretension to Imperial titles on the part of those who themselves were of forest origin. In other words, what anthropological/ethnographic publications on Indian society have attempted to tell us so far regarding the formation process of the major characteristics of the Indian society have been by and large ignored in historiography of Indian society in the early stages of its formation.[28]

I shall cite below three separate pieces of historical evidence in an attempt to bring out how the transformation of forest-space into a monarchical domain wrought irretrievable changes in forests and in the community of forest-dwellers.

The first piece of evidence dates to AD 582-9, well within the Gupta period, referring to *Parivrājaka* (literally, 'wandering') *mahārāja Samkṣobha*, who, in turn, mentions his chronological location in terms such as: 'in the enjoyment of sovereignty by the Gupta kings' (*Gupta-nṛpa-rājya-bhuktau*).

The ruler was governing the *Ḍabhālā-rājya* which was a part of eighteen forest kingdoms (*s-āshṭā-daś-āṭavi-rājy-ābhyantaram*) and which had come to him as inheritance. Apart from the brief reference to the Guptas and genealogy of the family, thus establishing the authenticity of the transition from *aṭavi* space into space for a *rājya*, a few elements in the contents of the record merit attention. First, like the Guptas, the *Parivrājaka* ruler too was a *paramabhāgavata* (great devotee of the *Bhāgavata* faith), the family having descended from *Maharṣi* king *Suśarman* 'who had learned the whole truth of the fourteen sections of science' [i.e., four Vedas, six *Vedāṅgas*, the *Purāṇas*, the *Mīmāṃsā*, the *Nyāya* and Dharma], in addition to being of *Bhāradvāja gotra* and comparable to Kapila.

The entire record in fact is a statement of the process of Sanskritization of a space which was, as the record itself suggests, a forest space (*aṭavi*). The practices which are highlighted through this copper-plate grant are: (i) assurance of constant engagement in establishing *Varṇāśrama-dharma*, (ii) gift of half of a village to the temple of *Piṣṭapurī* (mentioned also in other records from the area), a goddess probably originally of *aṭavi* origin, (iii) the use of Sanskrit for the record and appropriate citation from the *Mahābhārata* to ensure maintenance of the gift and to highlight merits of such gifts. All major ingredients which constitute *Brāhmaṇical* ideology in relation to royal power are present in the record; there is nothing of the world of the forest-dweller in it, except perhaps the goddess, metamorphosed and institutionalized through the *devakula*, the temple.

The background to the second piece of evidence that I cite is provided by, mainly, a type of historiography of post-Gupta India, particularly of peninsular India. In this historiography, the 'social distribution' of property and economic resources is admittedly largely neglected; it nevertheless highlights an important process: movement from wet-zone 'nuclear' political zones to dry zones which become new loci of agricultural expansion and enterprise, and of centres of new political elites[30]. Unfortunately, available historiography does not clearly tell us what the social origins of the new political elites were, particularly because by the time they became visible in written records, especially inscriptions, the actual origins have been concealed under a Sanskritic garb. Keeping this historiography in mind, I cite, as an illustration, extract from an inscription of the mighty *Hoysaḷas* of south Karnataka, who originated among the community of forested hill region of Chikmangalur and still claimed to be the leaders among hill-chiefs (*mālepas*), but in a completely Sanskritic/Brahmanic idiom:[31]

> The lineage of (that) *pratāpa-chakravartti* is as follows: *Brahmā* was born from the lotus-nave of Hari (*Vishṇu*), from him Atri, from whom moon, from him Budha, to whom *Purūrava*, from whom king *Āyu*, to whom Nahusha the crest-jewel of kings, to him the excellent *Yayāti*, from him Yadu, an abode of bravery, was *borń* and by whom the Yadu race became famous. The *Yādava* named *Saḷa became Hoysaḷa* when that fearless *Hoysaḷa* shot the tiger in the presence of goddess *Vāsantikā* of *Sáśákapura* on the excellent divine sage saying 'hit the tiger, *Saḷa*'.

The final piece of evidence that I cite comes from a late sixteenth century text from the western part of Bengal, the *Caṇḍimaṅgala* of *Kavikaṅkaṇa* Mukunda*rāma* Cakravarti.[32] This much-cited text in verse, one among several versions of the story, juxtaposes two narratives, one around a wealthy merchant householder and the other around a hunter who receives the favour of the goddess.

Appropriately, the locales of the two narratives combined in the text are different: a settlement mentioned both as a *grāma* and a *nagara* (urban centre) for the first, and initially a forest space for the protagonist *Kālaketu* for the second. However, *Kālaketu*, born as the son of Dharmaketu, a *vyādha* (hunter) and 'a chaser of deer' was really *Nīlāmbara*, son of Indra, born on earth as a hunter to initiate the *pūjā* of the goddess for eight consecutive days. *Kātaketu* grew up splendidly as a hunter, having been trained in all the required skills, and, in fact, when *Kālaketu* got married, his dowry consisted of a bow and three sharp arrows. And yet, even though there is a profusion of women of hunter-community (*vyādher nārī*), it is the Brahmin (*dvija*) who performs the ceremony in conformity with essentially non-tribal rituals. It is, therefore, in the preordained scheme of things that the forest-dwelling hunter, an incarnation of Indra's son, would, through the grace of the goddess, obtain wealth, cut down the forest, initiate the establishment of a town by settling varieties of social and professional groups (including daily-wagers) and having ruled in the town which is like *Ayodhyā-purī*, return to heaven after the expiry of his curse. All this because the goddess desired to be worshipped in the town:

Caṇḍikā says, *Kālu*, the son of a hunter,
Construct my abode in the middle of the town.

There are many meanings interwoven into the multilayered texture of the *Kālaketu*[33] story which is the preliminary part of the text *Caṇḍimaṅgala*. What seems most significant to me as pertaining to the theme of the paper is the use, by a rural poet, of the forest and the forest-dwelling hunter of a vaguely remembered memory as a background of a narrative which next moves through a completely different spatial location of towns, markets and kingdoms with its complex social organization and social/geographical network, ultimately to provide *Paurāṇic* legitimacy and universal acceptance of a goddess. There is thus a thread connecting the first piece of evidence with the evidence of *Caṇḍimaṅgala* in both of which Brahmanical elements take over.

Are such changes among forest-dwellers inevitable?

I close this paper by referring to a unique account, of life lived in the forest and among forest-dwellers in Bihar in the thirties of the twentieth century, by Bibhutibhushan Bandyopadhyaya, regretfully known to the general reading public only as the author of *Pather Panchali* (lit. 'Saga of the Road')[34]. The text I refer to is his *Āraṇyaka*. In the course of his life in the forest-space, Bibhutibhushan, out of his indomitable curiosity, sought out the habitat of a forest *rājā*, a respected leader of his forest-dwelling tribe but sharing his simple subsistence practices with fellow members of his community, and blissfully unaware, along with his innocent great grand-daughter, of the world outside his own forest-space. The tribal *rājā* was no monarch because he had no *brāhmaṇas* to give land to, no temples to construct, no courtiers and warriors to patronize and no genealogies fabricated to elevate him to the status of a *kṣatriya* which his successors could hang on to.

The ancestral shrine of this tribal *rājā*, rather remote and inaccessible, was not a pilgrimage site, to be visited regularly by members outside the community and sanctified by Sanstritic rituals. In other words, the tribal *rājā* was a poor tribal chief and his space an *araṇya*, not a *rājya* or kingdom, less than hundred years ago, in a locality which subsequently fast got caught up in the vortex of stratified social structure, religion and politics in contemporary times.

The argument of this paper is that with the emergence of the complex society, represented by the state with the twin requirements of cohesion and punitive action, the forest-dweller and his forest-space enters into the orbit of a different social order, the degree of the incorporation depending on a number of historical factors. When the forest itself becomes the locus of the state, through various processes of interaction with the complex society, the fundamental change in forest society is in the formation its own elite groups and complete reorientation in the nature of its relationship with the world outside, not only in wide-ranging changes in its landscape but in the acquisition of both symbols and substance of political authority of the contemporary complex society as well. This, in short, was the major mode of the formation of power elites in early Indian history, as I would consider viewing it.

NOTES

1. An interesting introduction to such perceptions is available in Romila Thapar, 'Perceiving the Forest: Early India', *Studies in History*, New Series, vol. 17. 1(2001), pp. 1-16.
2. See G.D. Sontheimer, 'The *vana* and the *kṣetra*: The tribal background to some famous cults', in G.C. Tripathi and H. Kulke, ed., *Religion and Society in Eastern India*, Delhi, 1994, pp. 117-164.
3. F. Zimmerman, *The Jungle and the Aroma of Meats*, Indian edition, Delhi, 1999, p. 3.

4. For relevant discussions see K. Sivathamby, 'Early South Indian Society and Economy', *Social Scientist*, no 29(1974), pp. 20-37; G.D. Sontheimer, *Pastoral Deities in Western India*, tr. Anne Feldhaus, New York-Oxford, 1998, chapter 2.
5. This is embedded in the theory of the origin of the State in the early Buddhist texts. See T.W. and C.A.F. Rhys Davids, *Dialogues of the Buddha: Translated from the Pāli of the Dīgha Nikāya*, part 3, London, 1921; R.S. Sharma, *Aspects of Political Ideas and Institutions in Ancient India*, third revised edition, Delhi, 1991, pp. 65-66.
6. R.S. Sharma, *Aspects of Political Ideas*, ch. 3. This is evident from the power of the king to wield *daṇḍa*.
7. *The Rig Veda: An Anthology* (Penguin Books India, 1994), pp. 242-3. The entire translation of the hymn has been cited here because it seems to have significant implications, not raised by O'Flaherty, for the theme of this paper.
8. The titles of texts suffixed with such titles as *āraṇyaka* and *upaniṣad* would suggest this. While *āraṇyaka* means 'relating to the forest', '*upaniṣad*' carries the sense of 'mystic knowledge or instruction'.
9. Patrick Olivelle, *Dharmasūtras*, p. 64.
10. Olivelle, *Dharmasūtras*, pp. 65-6.
11. While settlement units have been viewed from the perspective of administrative arrangement in the *Arthaśāstra*, more useful are references in early Buddhist texts to villages, commercial centres, cities, big cities and so on. For relevant discussion see N. Wagle, *Society at the time of the Buddha*, Bombay, 1966; A. Ghosh, *The City in Early Historical India*, Simla, 1972.
12. This authority would show why even to *Sārvabhauma* rulers the forest hermitages represented liminal space, demanding deference and observance of appropriate codes, symbolized by reining in of horses, getting down from chariots and, of course, suspending the royal pleasure of hunting. The most telling example from classical Sanskrit literature is provided by Act I of *Kālidāsa*'s *Abhijñāna-Śākuntalam*, starting, after the prologue, with a reference to the chase of a black antelope (*kṛṣṇasāra*).
13. See for example the *Arthaśāstra* of *Kauṭilya*. The *bhūmicchadravidhānam* section of *adhyakṣapracāra* (2.2) recommends, among other measures, that the king should establish on the frontier of his *janapada* elephant-forest (*hastivana*) to be protected by *aṭavi* and that *the adhyakṣa* of *nāga-vana* (elephant-forest) must ensure their protection and see to the elimination of those who kill elephants. 'Those dwellers of the forest (*vanacara*) who themselves collect the two tusks of a dead elephant may be rewarded with four and a quarter *paṇas*. There are many other details, including the relative qualities of elephants from different regions. See also *Arthaśāstra*, 2.31-32.
14. Romila Thapar, *Aśoka and the Decline of the Mauryas*, third revised edition, Delhi, 1997, p. 68 ff. For forest-dwellers in the Maurya period see also, Aloka Parasher-Sen, 'Of tribes, hunters and barbarians: Forest-dwellers in the Maurya period', *Studies in History*, vol. 14.2 (1998), pp. 173-91.
15. See the suggestion of Irfan Habib in his 'The Peasant in Indian History', incorporated in his *Essays in Indian History: Towards a Marxist Interpretation*, New Delhi, 1995, pp. 124-5.
16. *Arthaśāstra*, 2.1. The translation, although based on translations of R.G. Basak and R.P. Kangle, is slightly different. See R.G. Basak, *Kauṭilīyam Arthaśāstram*

(Sanskrit text and Bengali translation), vol. I, Calcutta, 1950, pp. 63-7; R.P. Kangle, *The Kauṭilīya Arthaśāstra*, pt 2 (English translation), Bombay, 1972, p.6.

17. Major Rock edict 13: for the text of the inscription see D.C. Sircar, *Select Inscriptions bearing on Indian History and Civilization*, vol. I, second edition, Calcutta, 1965, pp. 34-7.
18. Aloka Parasher-Sen, following J. Bloch, thinks that the relevant sentence means: 'They are even told that they repent and do not kill anymore', 'Of tribes, hunters and barbarians', p. 185. However, Sircar's Sanskrit rendering of the passage seemed preferable to me.
19. Major Rock Edict I of *Aśoka*, Sircar, *Select Inscriptions*, p. 15. *Samājas*, like *utsavas*, were community festivals which could be observed over days and nights, and could involve sacrifices, eating of meat and drinking liquor. It is interesting that some later rulers, even though Jaina by faith or extending patronage to Buddhism, actively sponsored such festivals.
20. See the translation in Romila Thapar, *Aśoka and the Decline of the Mauryas*, p. 264.
21. B.N. Mukherjee, *Studies in the Aramaic Edicts of Aśoka*, Calcutta, 1984, p. 33.
22. Translations of this and other verses of the dialogue between *Vālin* and *Rāma* are from Rosalind Lefeber, *The Rāmāyaṇa of Vālmīki*: An Epic of Ancient India, vol. 4 (*Kiṣkindhākāṇḍa*), Princeton University Press, New Jersey, 1994, pp. 87-95.
23. See reference to the implication of the Tipperah inscription of *Lokanātha* in my *Aspects of Rural Settlements and Rural Society*, p. 28.
24. For the text of Allahabad *praśasti*, eulogy of deceased king Samudragupta, see Sircar, *Select Inscriptions*, I, pp. 262-7.
25. For example, although the hill states of the Himalayan belt may trace, in their traditions, their antiquity far back in history, it was the *gaṇasaṃgha* form of polity which can be located along the Himalayan belt in some areas in the pre-Gupta period. For suggestion about the post-Gupta major transition see my paper 'Geographical Perspectives, Culture Change and Linkages: Some Reflections on Early Punjab', in *Studying Early India: Archaeology, Texts, and Historical Issues*, Delhi, 2003, pp. 48-65. For an example of the chronology and nature of an early Himalayan state, see D.N. Jha, 'State Formation in a Peripheral Region: The case of Early Medieval Chamba', in *The Feudal Order: State, Society and Ideology in Early Medieval India*, Delhi, 2000, pp. 197-210, edited by him.
26. See J.F. Fleet, 'Inscriptions of the Early Gupta Kings and their Successors', *Corpus Inscriptionum Indicarum*, vol. 3, reprinted by Indological Book House, Varanasi, 1970, pp. 112-16. For references to *aṭavi-rājyas* or, similarly, Pulinda-*rāja-raṣṭra* and significance of such references see my paper '"Reappearance" of the Goddess or the Brahmanical mode of appropriation: Some Early Epigraphic Evidence Bearing on Goddess Cults', in *Studying Early India*, pp. 172-90.
27. For the most effective statement of this argument, followed by other writers using it as a basic model, see the classic by R.S. Sharma, *Indian Feudalism*, A.D. 300-1200, Calcutta, 1965. See also the collection, D.N. Jha, ed. *The Feudal Order*.
28. An early major insight into the processes of passage from tribal to Hindu was provided by Nirmal Kumar Bose, *Hindu Samājer Gaḍan*, Calcutta, 1947. Bose illustrated these processes by taking up the tribal spaces through which Chaitanya

travelled to Nilacala or Puri. An English translation of this seminal work, by Andre Beteille, is available now as *The Structure of Hindu Society*.

29. Fleet, *Inscriptions of the Early Gupta Kings*, pp. 112-6. See, as a major secondary publication H. Kulke, *Kings and Cults: State Formation and Legitimation in India and Southeast Asia*, Delhi, 1993, *passim*.
30. Bibliographical references will be many, and I cite here only one reference which sums up the findings. Cynthia Talbot who has worked on Andhra between 1000-1500 writes: 'By 1200 the long-standing geo-political pattern in south India had been altered forever... Like the Kakatiyas who were based in the mixed zone Telengana, the other regional polities of the time were based outside the most cultivable lands... "a shift of dominance in peninsular politics from the old riverine core kingdoms of the earlier medieval age to the larger zone of upland, dry" territory occurred,' *Precolonial India in Practice: Society, Region and Identity in Medieval Andhra*, New Delhi, 2001, pp. 43-4.
31. *Epigraphia Carnatica*, new edition, vol. 7 (Mandya), Mysore, 1979, p. 565.
32. The edition that I have used is the one by Sukumar Sen, Sahitya Akademi, New Delhi, 1975. See pp. 28-105.
33. For the general pattern of use of the story of *Kālaketu* by historians who underline the rural, urban and commercial aspects of the story see Gautam Bhadra, *Mughal Yuge Kṛsi Arthanīti O Kṛṣak Vidroha* (in Bengali), Calcutta, 1983, pp. 56-7, 63-4, 127; Rajat Datta, 'From Medieval to Colonial: Markets, Territoriality and the Transition in Eighteenth Century Bengal', *The Medieval History Journal*, vol. 2, No. I (1999), pp. 145-6. Also David L. Curley, 'Kings and Commerce on an Agrarian Frontier: *Kālaketu*'s story in Mukunda's Candimangal', *Indian Economic and Social History Review*, vol. 38. 3(2001), pp. 299-324.
34. For the text of *Āraṇyaka* I have used the paperback edition, published by Mitra and Ghosh, 14th print, Calcutta, 1998, Chapter 11. The story is that of a remote forest domain of the forest-king Dobaru Panna Birabardi. He is mentioned as having been a leader of Santal rebellion against the British in 1862, was highly respected among the forest-dwellers and was extremely poor and found tending his cattle. Although believed to having been descended from a family of rulers of an extensive forest-space between Himalaya in the north, Chhota Nagpur in the south, Kosi river in the east and Munger in the west, Raja Dobaru Panna was not a ruler in the sense the King of Jeypore, the ruler of the 'jungle-kingdom' in Koraput, Orissa, was: B. Schnepel, *The Jungle Kings: Ethnohistorical Aspects of Politics and Ritual in Orissa*, Delhi, 2002.

Re-thinking Forest, Forest Dwellers and Ecological History

K.S. Singh

I compliment the Indian History Congress and the Forest Department of Madhya Pradesh government on organizing a Panel Discussion on the theme, 'Tribes, Forests, and Social Formation in History' at an opportune moment. An ecological history in India is taking shape, with contributions from ecologists, anthropologists, sociologists and some historians. However, an ecological history of India is still to be written comprehensively by historians using their craft to relate and correlate information and knowledge about ecology and change to the larger processes of history, in a holistic framework. An interesting discourse is going on about the relationship of humankind and forest, relationship of forest and agriculture, gender dimensions particularly the role of women as gatherers and hunters and so on. We will discuss some of the issues in response to the note circulated with the letter from the Indian History Congress.

Various terms have been in use for forests in our classic and folk, particularly tribal literature. The word *jangal* (or its Anglicised form, jungle) seems to be quite old and still in use, followed by terms like *aatavika*, *mahakantar*, *jharkhand* in Sanskrit and Persian, *bir* in Mundari, etc. A distinction is being made between *jangala* and jungle, the first applies to the forests in the dry land of the western part of the subcontinent which was the seat of the early civilization, of the Vedas, the sacred land, while the second applies to the dense forest of central and eastern India. The word *jharkhand* generally means a forest tract and as such it is mentioned in Sanskrit texts. It also means specifically a forest region of eastern India: the term 'Dakshina Jharkhand' occurs first in Kendrapara (Orissa) copperplate grant of the thirteenth century. The term also occurs in western Orissa, Greater Jharkhand and the present-day Jharkhand and is mentioned in medieval Sanskrit and Persian texts. A Sanskrit word, it is adopted by the tribes speaking Austric and Dravidian languages and emerges as a powerful symbol of tribal identity in the 1940s and 1950s and lends its name to a state established on 15 November 2000 through a process of cultural osmosis and political mobilization.

There is no term for tribe in ancient literature. There are words for forest dwellers in both Sanskrit and *bhasa* literatures. The tribe is an Anglo-saxon word, a colonial concept. A tribe officially notified as a scheduled tribe today lives in deserts, riverbelts, the plains , the islands, in relatively isolated habitats but mostly in forests. The tribes of today and the forest dwellers of yesterday are almost similar. We have today a profile of 3,000 communities including tribes which share a good many features such as endogamy, commensality, hierarchy, exclusion, etc., and a sense of identity. There are differences between communities and between regions. The tribal communities spread all over the countries are marked inter alia by relative isolation and distance, cohesiveness, egalitarianism, incipient forms of hierarchy, rapport with environment, subsistence economy with simple tools of production, which is being radically transformed.

There is a difference in perception of forest as a universe between the tribals and non-tribals and between the classical/textual traditions and the folk and oral traditions. For example, we have this perception of forest given by Dr Suniti Kumar Chatterjee in an explanatory note in the translation of *Varna Ratnakar, a 14th century Mithila classic.*

> There is a mystery also in the forest—one finds not only the magic of the hunters to catch their quarry, but also one hears the song of the Kinnaras, and the airs sung by the Vidyadharas, celestial beings: the forest indeed is the abode of horror, and also of beauty. This *kallol* finally closes with an account of a sage's hermitage. (Chatterjee et al., 1940)

In contrast to the Sanskritic view of forest as mysterious, and a poet's description of forest as 'dark and deep' is the tribals' view of forest as their world, the abode of their spirits, the seat of their culture, the source of their living and livelihood. The Abujhmariah call their land, *sringarvelu dweep*, the land of colourful abundance, and all forest regions in tribal folklores are described as enchantingly green.

Forests have always been recognized as an important and distinct resource region, a source of food, goods and assets, such as elephants, minerals and timber. In lexicographical works such as the *Mahabharat* and *Varna Ratnakar*, forests are listed as a resource or a resource region along with mountains, rivers, pastures, deserts, etc. The people or *janah* inhabiting these resources regions are also mentioned. The forest and forest dwellers (such as *jangalah*, *dandah*, etc.) in the *Mahabharat* constitute a large number of entries in the list of peoples in the epic. In *Varna Ratnakar* the forest dwellers and hill dwellers account for 17 and 10 communities respectively.

Continuum versus Dichotomy. Agriculture and forest have always being treated as dichotomous categories. Recent studies highlight interdependence and continuities. Forest has been associated with the early life of humankind where they used forest resources for a living, where they domesticated plants and innovated early forms of agriculture and gradually moved away from gathering and hunting to advanced and primary modes of subsistence. However, there was no marked dichotomy between forest and agriculture as perceived later. There always existed a continuum between forest and agriculture and it still exists in three ways.

First, the tribes developed their technology to exploit their resources and internally managed transition to rudimentary forms of agriculture such as shifting cultivation, terrace cultivation and some forms of even settled cultivation. The studies of such practices that survive today, and which are even flourishing, suggest an extraordinary knowledge of the environment, of plants and crops, a sensitivity to resources like soil and water, and the practice of the earliest techniques of conservation of soil and water through bunding, water harvesting, construction of small dams, even wells, etc. There are tribal myths that describe the beginning of endogenous agricultural transformation with the use of iron, knowledge of crops, etc.

Second, the tribes were always and still are the first, probably the greatest reclaimers of land, doing all the arduous chores of breaking soil, clearing forest, and removing vegetation. The process of reclaiming land from forest for cultivation has been described by Birsa Munda in the following words:

> This was the period [*satyuga*] when their ancestors reclaimed the land from the forests infested with wild animals. Lions roared, snakes hissed and elephants trumpeted. They lived in caves and made friends with the wild

animals, organized themselves, and cleared the forests. They founded villages, constructed embankments, tapped natural springs, planted trees and orchards, dug well and tanks. They started cultivation to earn their living. Thus did they spend many years in happiness. They did not kill their brethren and relatives. They discriminated and did not make jokes with anybody's sister or mother. They lived honestly and together. If the enemy put one to trouble, all came together and drove him away. Thus did the *satyuga* pass away. (K.S. Singh, *Birsa Munda*, p. 164.)

Third, the tribes have always been and are still part peasant and part tribe or part forest dwellers. Forests are still part of life of most of them, of their economy, their culture, their belief system. This is true of even that phase of history when reclamation of land was taken up as an instrument of state policy and when advanced peasant groups started moving into tribal areas in search of land. The role of these peasant groups in disseminating advanced technology of agriculture, the concept of agriculture season, knowledge of crops are well documented for some of the regions of India. All this led to further peasanting the tribals and bringing tribals and non-tribals closer economically and culturally. However, the tribals' links with forest were never lost, totally and irretrievably.

The continuum is also suggested by the knowledge system and practices adopted by peasants in various regions. An extraordinary range of knowledge of local varieties of paddy, pulses, roots and tubers, vegetables, etc., exists in all regions. Planting of trees (particularly fruit-bearing trees) and protection of sacred trees are considered a *sanskara,* for a peasant, an act of merit, described in *dharmshastras* as more meritorious than even begetting progeny. Water harvesting practices are continued and developed, as is knowledge of local varieties of various crops, soil and nutrients. A regionwise study will show the extent of the continuance of tribal or early practices in later day agriculture. There has been some study of this in Chotanagpur in Jharkhand, which suggests existence of different layers, one upon another, of agricultural practices and their interpenetration.

As mentioned earlier, the tribes had and still have natural rapport with their environment. Even in reclamation of land, a tribal shows sensitivity and tenderness which is uniquely his own. He would not destroy a tender plant, he would not fell a fruit-bearing tree or trees considered sacred, he would not take for a toothbrush a twig from a young and tender sapling but from an old and twisted tree. Traditionally they have never exploited or over-exploited forest resources and have always made a judicious use of or have been economical in the use of the resources. But while they have conserved and protected their environment including forest, they have not planted forest—there was no need for it as forests were abundant. In fact it has been said that multiplication of forest has been more a function of moths and butterflies. Therefore, there has been a radical change in recent years as tribals particularly tribal women have been emerging as planters of trees and their protectors, and as a new culture of forest development and tree plantation has begun making an impact. There are also instances of tribals protecting forests resulting in self-generation or re-generation of forest.

Conflicts, Struggles and Integration. There were hostilities between tribes and non-tribes, the latter representing the peasants, pastoralists, etc. There were conflicts also derived from different modes of living, and systems of beliefs, particularly between *yajna* performing peoples and the autochthons. The mythological accounts of these conflicts contained in the Puranas—and now being presented garishly over various TV channels—demonstrate the roots and dimensions of the conflict between the forest dwelling people stereotyped as demons/ *asuras/rakshsas*, on the one hand, and the rich, the powerful and the *devas*, on the other. The dialogues suggest ethnic, cultural, class and even sibling rivalries between the two groups. One can also discern the working of the processes of integration and acculturation in terms of accounts of common origin and sharing of elements of culture between the two groups. The mythologies have probably no basis in history, although Dalit historiography makes use of them. They only serve to explain the rise of and legitimize the mother goddess cults in different parts of the country.

There are derisive references to the tribes in early literature. The Puranic stereotypes of tribes as ugly, savage, living outside the pale of civilization survive in medieval accounts of scholars like Abul Fazl, whose description of the Gonds reminds one of the Puranic description of Vena. The Gonds were no savages; they founded states, the last of which survive in Bhopal.

As the pastoralist Aryans set up *ashrams* in forest, these became the meeting ground of all communities, a centre of acculturation. As

Valmiki tells us, 'There [in the hermitage of Agastya] the Devas, the Yakshas, the nagas, and the Bird-like community live together, restricting their food habit for the sake of their cultivation of religion' (Singh, *The Scheduled Tribes*).

Thus we have legends of friendship and kinship cutting across communities, a sense of equality and mutual respect. To mention a few instances, Dasharath and Jatayu, Ram and Guha, Pandu and Bhagadatta were products of *ashram* schools.

The Brahmans have been described as agents of acculturation by Kosambi, Ghurye et al., and of diffusion of technology by historians dealing with the impact of land grants in remote regions. It should be noted that they left local customs largely untouched. Our studies show that there were other agents of change, such as the peasant castes and artisans moving into forest areas who were instruments of dissemination of technology and of Puranic forms of Hinduism and Bhakti traditions among tribes (Singh, *Tribal Society*).

Therefore, there is perceptible change in the attitude towards the tribes in later classical literatures which describe the forest dwelling communities in terms of respect for their valour, their prowess and their achievements. Thus the Kiratas and the Savaras are described in glowing terms. It is therefore not correct to say that all tribes became Shudras. Many tribes both in the north-east and middle India founded states, acquired political power, and claimed the Kshatriya status while retaining their *jati*. Their descendant groups still enjoy high status as former zamindars, jagirdars, talukdars, etc. But the majority of the people were no doubt reduced to the ranks of workers and labourers who provided free or cheap labour; others known for their skills became the caste groups performing specialized services.

There is no linear or unidimensional process of the transformation of tribes into peasants. Today we see diversities, variations, continuity and autonomy in all social formations. Assimilation was never complete, and acculturation is an ongoing process. Therefore, the old theory of assimilation or absorption needs to be modified as information about various levels of identity, degrees of autonomy, patterns of interaction becomes available. The profile of tribes or forest tribes or castes and of their relationship that is emerging is heterogeneous.

The Gender Dimension: Gathering versus *Hunting*. The forest dwellers, both men and women were hunters, who hunted for food and for sport occasionally, but the representatives of the state, ancient, medieval and colonial, brought up on the ideology and traditions of *shikar,* hunted wildlife, as a mark of status, prowess, manliness, even annihilated them—each succeeding regime got worse when it came to indulgence in such massacres occasionally—until the notion of conservation emerged towards the end of the nineteenth century. There were however cases of conversion in the case of Ashoka who prohibited killing of some animals and birds and in the case of Emperor Akbar, who forbade killing of wildlife, after grisly experiences.

A debate is going on about the hunting role of women which was later usurped by men. The Indian material, historical, anthropological and mythical describe women as huntress. The earliest evidence of this is at Burzhom, a neolithic site in Kashmir valley where a woman along with a man is shown spearing a large mammal. We have descriptions of women shooting flying birds in sky in Vedic literature and killing and carrying animals in Rajasthan (Jaiswal, 2001). In the medieval period, Durgawati and Noorjahan shot tigers. There never was a ban on hunting by women except during the period of pollution. The Chotanagpur tribals still celebrate a ritual hunt called *jani shikar* every twelve years when they dress like men, wearing turbans and carrying hunting equipments, go out to hunt minor animals, and now with restrictions imposed on hunting, they hunt domesticated animals and birds. These historical images of women have been translated into the Puranic and mythological accounts of mother goddesses, taming, riding and killing major mammals. However, there has been a gradual almost unexplained surrender of women's role in hunting to men, as in agriculture and many other fields, under the influence of patriarchal norms which have penetrated tribal society. All female hunting goddesses barring one or two have faded away: Shiva has emerged as the archetypal hunter, commemorated in Bharavi's classic *Kiratarjuniya Mahakavayam.* Khandoba was originally the god of hunters, pastoralists and shepherds (Sontheimer, 1997).

However, unlike the huntress, women's role as gatherers shows a remarkable continuity. From gathering roots and tubers and other forest produce to maintaining family in the early period, women today apart from performing this role also collect fuel-wood and fetch water to keep hearth burning and play a considerable role in gathering and marketing what is called non-timber produce or minor forest produce.

There is an unbroken continuity in this role which explains why tribal women have better access to food, drinks and earnings, and why they are relatively free and autonomous. Interestingly, while man once was the woodcutter, going around with his *dao* or pick axe, cutting branches or felling trees, women now along with men are playing a major role in planting trees and protecting forest, fostering a new culture of tree planting by tribals, as mentioned above. Women are the major beneficiaries of various experiments in reforestation.

Colonialsim, Forest and Forest Dwellers. The pre-colonial state promoted extension of cultivation which resulted in shrinkage of forest. The colonial period witnessed radical changes in the very concept of ownership of forest and its management. While all states of India, ancient and medieval, claimed ownership of all resources, they in practice left local communities largely in control of their resources. The colonial regime not only claimed but enforced its ownership of resources; thus forest became for the first time a resource controlled and regulated for the benefit of the British empire and British capital. This brought the colonial regime in direct conflict with forest dwellers or the tribes with deeply entrenched notion of ownership of forest and its resources. And so we have for the first time a series of forest-based protests, uprisings and movements in the colonial period compared with the pre-colonial when, because of relative abundance of the forest, instances of such reaction against the state were probably fewer.

Second, the colonial government inspired by the Victorian concept of the honest toiler (Sundar, 1997, p.120) and pressured by the peasant groups vigorously pursued a policy that promoted peasantization of tribals. There was an enormous extension of cultivation by peasants and tribals as successive settlement reports of tribal district show, though a large tribal area also remained unsurveyed in Madhya Predesh, Orissa and Andhra.

Third, the tribal practices such as *jhum* cultivation were considered harmful for commercial forestry and were discouraged. *Jhum* disappeared from large parts of tribal middle India and tribals mostly settled down to agriculture, though at a subsistence level and many of them became landless labourers. However, a healthy understanding of *jhum* cultivation emerged later as it began to be seen as a strategy of survival and subsistence for tribals and also as a practice that preserved environment in the long run.

Fourth, the colonial regime extinguished all local rights in reserve forests and instituted the system of direct management of forests. Local rights were recognized in private forest and later even in the reserve forest, but the conflict between the forest dwellers and the forest management began, the first continuing to perceive the forest as their resource and the latter resisting any attempt to encroach on the forest beyond the recognized rights. However, there were certain interesting experiments in local management of resources through *van panchayat* and lineage owned (*khuntkati*) forest. There were diversities of opinions and perceptions in the very interesting debate on the rights of people, ownership of local communities, model of management—some of them strikingly modern and even contemporary—in the run up to the Forest Acts.

While in the long pre-colonial period there were conflicts between the state, peasants and pastoralists on the one hand, and the forest dwellers on the other, it was during the colonial period, particularly from later half of the nineteenth century, when the commercial exploitation of forest intensified, that the forest based protests, uprisings and movements began on a larger scale. The Rampa rebellion of 1879-80 was directed against the restrictions imposed on shifting cultivation. Interestingly, Birsa Munda before he emerged as the prophet led the protest against the forest settlement operations that curtailed the rights of the people, such as the right to free fuel, grazing, etc., and demanded remission of forest dues. The *bhumakal* of 1910 in Bastar started as a protest against reservation. The Indian National Congress mobilized the tribals through a series of forest *satyagrahs* on forest related issues in the Central Provinces and Bombay Presidency. Kumra Bhimu led the greatest uprising of all in 1940 at Babejheri. In the north-east Sonaram R. Sangma (1869-1960) demanded a stop to ever expanding reserve forest in the Garo Hills.

Anthropologist Verrier Elwin criticized the Congress and the Gandhians for introducing plough cultivation, khadi and prohibition in tribal area. However, plough cultivation was being encouraged by the colonial regime as part of its strategy to reclaim tribals to civilization. The Congress was no doubt pro-*kisan*, but not at the cost of tribals in later stages when Gandhians took up their cause in the 1930s, and when the Mahatma endorsed their programme and called the tribals the last but not the least of India's children in the early 1940s.

Post-colonial Situation. The post-colonial situation has witnessed large-scale destruction of forest and wildlife, escalation of tension between forest management and forest dwelling communities, the enhanced role of timber mafia acting in collusion with a section of the tribal people and the bureaucracy, increasing incidence of protest and violence and movements to project and protect people's rights. After independence the Kherwars of Palamau under the leadership of Fetal Singh and the Kherwars of the adjoining area of Madhya Pradesh organized the first and the most articulate forest-based agitation in the mid-1950s. They demanded full freedom to operate in the forest, and an end to the restrictions imposed on their customary rights in forest. In all such movements we find traces of the primeval notion of tribal ownership of forest which is even now a burning issue. The post-independence scenario, particularly the last three decades is also marked by people's, particularly women's, participation in movements to protect forest and expand its coverage, however the impact of such a movement is limited.

There are other signs of hope emanating from the international movement for environmental conservation: the growing consciousness that forest dwellers have been the custodians of biodiversity and have an unrivalled knowledge of plant genes and thus hold the key to human development, the understanding that forest cannot be protected against the people and that people must have a share in the management of forest resources. A number of experiments are underway all over the country to institutionalize people's participation. There are many success stories from West Bengal and Andhra, Madhya Pradesh and Maharashtra. Punjab and Haryana have succeeded in combining the green revolution with a great deal of greening of the countryside. There is also a greater understanding of interdependence of various resource regions, of forest and agriculture, and a new concept of ecological services that underlines the need for compensating the forest dwellers and forest managers for the services they perform in conserving moisture, storing water, providing biomass, recharging groundwater, etc., for the benefit of the communities living in the plains or elsewhere.

Conclusion. The Indian History Congress has played a glorious role in writing the scientific history of our people. Ecological history is now emerging as a component of scientific history. It is the new frontier that is going to be comprehensively explored in the new century.

Ecological security is now increasingly seen as an essential part of human development. Therefore, historians have a social responsibility too. While the current writings generally give us a view from above, we require regional, micro-level studies in ecological history based on what Professor Damodar Dharmanand Kosambi described as combined methodology, drawing inputs from allied disciplines and drawing on an intensive field study of the people and their folklores. It is a challenge that should be taken up by young historians.

REFERENCES

Chatterji, Suniti Kumar and Babua Mishra (eds.), *Varna Ratnakar of Joytirisvara-Kavisekharacarya*, Bibliotheca Indica No. 262 (Royal Asiatic Society of Bengal, Calcutta, 1948).

Guha, Sumit, *Environment and Ethnicity in India, 1200-1991* (Cambridge University Press, Cambridge, 1999).

Jaiswal, Suvira, 'Female Images in the Arthashastra of Kautilay', *Social Scientist*, Vol. 29, No. 34, March-April 2001.

Kosambi, D.D., *The Culture & Civilisation of Ancient India in Historical Outline* (Routledge and Kegan Paul, London, 1965).

Singh, K.S., *Birsa Munda and His Movement, 1874-1901* (Oxford University Press, 1983).

——— (ed.), *Economies of the Tribes and their Transformation* (Concept Publishing Company, 1982).

———, *Tribal Society in India: An Anthropo-Historical Perspective* (Manohar, 1985).

———, *The Scheduled Tribes* (Oxford University Press, 1994).

———, 'Gender Roles in History: Women as Huntress', *Gender Technology and Development,* 5(1), 2001, Asian Institute of Technology, Bangkok (Sage Publication, New Delhi).

———, '*Varna Ratnakar* and its Ethnograpy', in Hetukar Jha (ed.), *Perspectives on Indian Society and History: A Critique* (Manohar, 2002).

———, 'The Ethnography of the *Mahabharata* and the North-East' (IIAS, Shimla),

———, 'Colonialism and Tribal Economy' (mimeo).

———, 'Changing Attitude to Forest and Nature: A Historical Review with Focus on Jharkhand' (mimeo), AIT, Bangkok.

———, 'Shamanism, Witchcraft and the Position of Women in Tribal Society' (mimeo), AIT, Bangkok.

———, 'Tribal Women: Resurrection, Demystification, and Gender Struggle' (mimeo), ICHR.

Sontheimer, Gunter-Dietz, *King of Hunters, Warriors and Shepherds: Essays on Khandoba*, ed. Anne Feldhaus, Aditya Malik and Heidrun Bruckner (Manohar, New Delhi, 1997).

Sundar, Nandini, *Subalterns and Sovereigns: An Anthropological History of Bastar 1854-1996* (Oxford University Press, New Delhi, 1997).

From Tribe to Caste? The Koragas of South Canara

Kesavan Veluthat

Historians of nearly all persuasions recognize that a transformation of 'tribes' into 'castes' has been an important aspect of the historical process in India. However, their studies hardly raise and answer the question as to where tribe ends and a caste begins. There are many problems in addressing it. For one thing, the use of terminology in most writings has been somewhat crude, with the result that terms such as 'tribe', 'caste', etc., are used without the precision and rigour expected in such discussions. Secondly, statements about the transformation of a tribe into a caste, granting an invariably higher position to 'caste', implicitly assume a hierarchy, as if the transformation was a natural process of promotion or graduation, as it were. The extent to which such assumptions are valid is seldom the concern in these writings, nor is the context in which categories such as these were originally constructed. It seems that nearly everybody repeats a statement, made by an authority once upon a time, with or without sufficient premeditation, often not considering the load that it carries with it.

In this paper, I intend to examine statements of this and a similar nature made about the Koragas, who have been officially recognized as a 'primitive tribe' by the governments of both Karnataka and Kerala.[1] They live in the Udupi and Dakshina Kannada districts of Karnataka and the Kasaragod District of Kerala, which region formed part of the South Canara District of the erstwhile Madras Presidency and where the Tulu language is spoken. There is great ambiguity in the way in which the literature describes and classifies the Koragas. Francis Buchanan saw in them a 'good-looking' people, although they were so many 'savages'.[2] They constituted a 'wild race' according to the Census of 1871 and 'aborigines' according to the next Census ten years later. In both, however, they are treated as a 'sub-caste' under the broader head of 'pariahs'. Edward Balfour described them as a 'servile race'[3] while M.T. Wallhouse is quoted to have thought of them as 'a very quiet, inoffensive race'.[4] It is obvious that Balfour and Wallhouse were using the expression 'race' in the same crude sense in which the term was used in the middle of the nineteenth century, without the anthropological rigour that came to be attached to it in subsequent decades. Gustav Oppert, however, used the term 'race' in describing the Koragas differently, when he equated them with the 'Gaudian race', likening them to the Kodagas (Coorgs) and speculating about the possibility of their being connected to the Kushanas![5] As per the Census of 1891, the Koragas were classified as a 'sub-caste' under the main head 'Forest and Hill Tribes'.[6] In 1893, the *Administrative Reports of the Madras Presidency* classified the Koragas as a 'praedial tribe' and 'a low set of people', the latter description deriving from an etymology of the word *koraga* from '*kora*', which is thought to mean 'defective, narrow, sharp, rude . . .'.[7] By the time we come to the next Census about a decade later, the Koragas find a place in the report as a 'wild tribe' of basket makers, which description Edgar Thurston accepts in his 'alphabetical empiricism' in relation to the *Castes and Tribes of Southern India*.[8] The Census of 1931 put them down as 'a caste which pollutes without touching' and, thereafter, census reports do not return such details. B.A. Saletore, the historian, showered praises on the Koragas who belonged 'to the larger group of wild tribes', who were both 'noted for their integrity and honesty' and 'addicted to warfare',[9] whatever that might mean. Purushothama Bilimale, writing in the 1990s, discusses the 'culture of the Koragas' in a monograph, where there is a whole section under the sub-heading 'from tribe to caste'.[10] Bilimale states that 'the Koragas, who have opened themselves up to modern education, are now turning towards a caste society'.[11] The assumption here is that they have left their tribal stage, but not quite arrived as a caste.

It is not just in modern literature that we find this ambiguity about the status of the Koragas in the social ladder. One point that no scholar writing on the Koragas has left out is a story of Hubbashika, a Koraga 'king' of old, who is said to have ruled the region of South Canara.

Typically, the story is in the usual structure where it was they who were the masters of this land and that it was following a misfortune, this time a treacherous act of the Brahmanical upper castes and their representatives, that they became such a lowly people. The account finds its expression, curiously, not only in the traditions of the Koragas themselves, but also of the Brahmanas. There are, of course, some variations in the two versions as also there are some local variations among the traditions cherished by the Koragas themselves.

Bilimale has recorded a few Koraga legends related to this theme. According to one, Hubbashika was the chief of the Koragas. He defeated Mayuravarman in a battle, but Lokaditya, the nephew of Mayuravarman, avenged the defeat in a fierce battle fought near Manjeshwara and drove him to the forests. The Koragas thus became a forest people.[12] Another version has it that Hubbashika came down the Western Ghats and gathered an army, which included Koragas. The brave Koragas set off to Mangalore, where Hubbashika fell seriously ill and was bitten by ants all over. He fled Mangalore for Manjeshwara where he defeated Angaravarma, son of Viravarma. He ruled over the region for the next twelve years.[13] The version, however, does not speak of a defeat and thus does not explain the lowly position of the Koragas. In yet another version, Hubbashika was a courageous and heroic king of Koragas, who got together an army of his people and went up to the Jain Angaravarma to ask his daughter's hand in marriage. Finding it impossible to defeat the arrogant but powerful Hubbashika in a straight battle, Angaravarma hatched a plan. He spread the entire yard with gooseberries, concealed by a carpet. As the Koragas came to the wedding reception, they slipped down and fell. Angaravarma's people butchered all those who fell down.[14] In still another version, Angaravarma of Manjeshwara defeated Hubbashika and his army in battle. Moreover, he harassed them so much that they finally submitted themselves to slavery. Many were stripped naked and driven to the forests. The Koragas who so fled to the forests started wearing leaves to cover their private parts.[15] In a slightly different version, it was Hubbashika who defeated Angaravarma. The defeated Angaravarma killed Hubbashika through black magic and witchcraft.[16] Bilimale has reported a story which seeks to explain the practice of Koraga women covering their modesty by leaves. According to it, there was a Koraga king by name Mutthale Arasu ('King with an Areca Spathe for a Headgear'), who sought an upper caste woman in marriage. Enraged by this audacity, men of the upper castes ordered that Koraga women should not wear any clothes any more. So they began wearing leaves to cover their modesty.[17]

It is clear that all these stories seek to present memories about a primordial Garden of Eden and a subsequent fall of the Koragas. There is, however, reason to believe that these memories are more constructed than remembered. One comes across the same theme, if in a slightly altered form, in the Brahmanical traditions as well. There is a narrative called the *Gramapaddhati*, which seeks to record the story of the creation of the land of coastal Karnataka and the establishment of Brahmanical settlements there.[18] The Hubbashika story is treated somewhat elaborately in this text. It is significant that the episode forms part of a series of 'disturbances' that the Brahmanical order in coastal Karnataka had to face, and as such it is as if the 'event' is presented as one of those obstacles which the Brahmanical world encountered, and surmounted, in South Canara. It is necessary to place the Hubbashika story in its context in the Brahmanical traditions for us to appreciate its significance in its fullness.

After Parasurama created the land, he looked around for Brahmanas to gift it. Finding none, he created a few by transforming local fishermen. He gifted the land to them and made all necessary arrangements for their comfortable life. Thereafter he left for mount Mahendra to perform his penance, assuring the Brahmanas that he would appear before them, should they need his help. One day they called out Parasurama as a test case. Parasurama promptly appeared; but, learning that they were merely testing him, he was livid with rage. He cursed them, divested them of all the properties and privileges he had conferred on them. They would be eating dirty food and wearing only rags thenceforth and would have to toil for a living as labourers in those very lands which were their own till a moment ago.

After some time, when the Kadambas came to power in the region, the rulers realized that theirs was a land without Brahmanas and hence undesirable. Mayuravarman, the founder of the dynasty, himself went to Ahicchathra and brought Brahmanas to the Canarese coast where he established them in different settlements. At this point, the fallen Brahmanas, now untouchable, protested saying that the land was actually theirs and that if Mayuravarman were to give it to the

Brahmanas, they had no place to go. So, they would not allow it; they would even beat the new Brahmanas up. Mayuravarman, armed with state power as he was, could have dealt with them the way he wanted. But he took pity on their poverty, as he was very generously disposed. Pleased by such of them who were still living in marginally decent ways, he decided to establish these sinful villains in the interstices of the Brahmanical settlements. They were not allowed any transaction whatever with the rest of the population; they could live as so many untouchable Sudras as labourers in the land now given to the Brahmanas, doing lowly work. Thus the land was once again worthy of Brahmanas.

Several generations later, when the Kadamba prince, Lokaditya, was still an infant, Hubbashika, a Candala, gathered several of his kinsmen around him. He harassed the people in various ways with a view to imprisoning Lokaditya after robbing him of all his wealth. Unable to put up with it, the queen mother fled with the child. Hubbashika took advantage of this situation and ruled the country in a most mean way. He plundered the wealth of the Brahmanas, both natives and passers-by. Unable to stand it, most of the Brahmanas returned to Ahicchatra. A few of them, however, stayed back on the plea that they could not leave their motherland (*matrsthala*). They went up to Hubbashika and promised him their support. The old travesties of Brahmanas, whom Parasurama had cursed, too declared their support to him. For, hadn't he retrieved from the interlopers of Ahicchatra Brahmanas the land that they had lost! So also, other non-Brahmanas of the land allied with Hubbashika and appropriated all land including areca and coconut gardens. Hubbashika, the Candala, ruled the country in a most wicked way, perpetrating untold miseries on the population.

Around this time, a man called Candasena went to Gokarna with the intention of giving his sister Kanakavati in marriage to Lokaditya. There he learned all about the misfortunes of Lokaditya and the bad deeds of Hubbashika. He thought it would be tactically wiser to feign friendship with the latter. Hubbashika was willing on condition that Candasena gave Kanakavati in marriage to him. Or else, he would kill Candasena and take Kanakavati by force. Candasena agreed to the marriage. A huge wedding hall was erected. Hubbashika and his Candala men were properly seated in the hall, of course with no arms on them. As already arranged by Candasena, Lokaditya, who was by now a grown up man, surprised the unarmed Candalas and pounced on them at the head of a huge army. Hubbashika was done away with and so were the many of his men. Those of them who survived, including their women and children, were put to bondage. They were assigned to agricultural labour and redesignated as several Holeya *jatis* such as Bakkada, Bakuda, Meraru, Goddianavaru, Bileyavaru, Halasavaru, Koragavaru and so on. They were to get only subsistence wages; they were liable to be beaten up every now and then. These 'slaves' would be just so many items of property, with no other rights than to livelihood. The happy ending of the story is the return, if after a lot of persuasion of the Brahmanas who had gone back to Ahicchatra and their resettlement in the region.[19]

There are many interesting features here. To begin with, what is interesting is the way in which both the Koraga and Brahmanical traditions match in essential details. It is but natural that Hubbashika is a hero in the Koraga version and a villain in the Brahmanical version. That he ruled over the region and that it was by an act of treachery that he was got rid of is a common theme in both the traditions. While he was a Koraga king in the Koraga tradition, he was just a Candala, a name used in Brahmanical texts to denote any one of the numerous forest-dwellers or 'primitive peoples', described as resulting from a *pratiloma* union of a Brahmana female and a Sudra male. The Koragas are just one among those groups who were reduced to slavery following the defeat of Hubbashika at the hands of the treacherous army of the Brahmanized rulers of the Kadamba family. They were among the 'Holeya *jatis*' of South Canara such as the Bakkada, Bakuda, Meraru, Goddianavaru, Bileyavaru, Halasavaru and so on. There is nothing exclusive about the Koragas in this version, while Hubbashika is exclusively a Koraga hero in their version. Nor are the other castes mentioned known to cherish the Hubbashika story.

The occurrence of similar accounts in versions of parties thought to be antagonistic in their interests is often taken as corroboration of the historicity of the event. What such a convention in historiography has done in this case is to offer the necessary scholarly debates in relation only to the spatio-temporal coordinates of the hero/villain and his men. Thus, scholars date Hubbashika variously to 1440 BC, 900 BC, 447 BC, 'between the second and fifth century AD', and sixth century AD.[20] On the subject of his home, too, there are equally wide speculations, ranging from Ethiopia to Central Asia to Ananthapur in Andhra

Pradesh to Ananthapura in the Shimoga District, next door to South Canara. The Koragas are accordingly thought to be descendants the Kushanas (doesn't the name Hubbashika rhyme enough with the Kushana name Huvishka?[21]); of Ethiopians (not only are there a people described as Habashat but there are also imagined similarities between the Nagari script and Ethiopian, and between the architecture and sculpture of India, Egypt and the neighbouring countries![22]) and people of Rayalaseema and the Malnadu regions. All these debates, however, operate within the same assumption, offering variations just in the narrative. The historicity of Hubbashika and the event which he is associated with is not contested in this fashion of historiography.

It is important to remember that the function of these stories, historical as they are, is not to keep a record of 'what actually happened'. It may not be worthwhile to apply the Rankean method to these statements for they have functions different from those of what are conventionally thought to be 'sources' of history. It is significant that in the Koraga traditions, Hubbashika was their chief, ruling the land once upon a time, when they enjoyed everything that was good. It was following an act of perfidy that they lost out, that they had to flee to the forests, that they had to wear only leaves or rags to cover their modesty. Such nostalgia for a lost golden age, out of which one was cheated, is not only a wish-fulfilment for those who, smarting under the degradation of subjection, cherish it but also an expression of protest against the existing order of things. Such stories, which present the depressed classes as once-upon-a-time-masters, are numerous in India.

In fact, the origin myths which the Koragas cherish are comparable. According to one, nothing existed in the beginning, not even the earth. Everything was submerged in water. There were only two living beings in that vast expanse of water—a 'male' and a 'female'. They were floating about in a vessel of a *dolu* drum when god Narayana appeared before them and asked them if they wanted anything. They answered in the negative and Narayana disappeared. After a few days Narayana came back and asked the same question, only to get the same reply. On the third occasion, not able to resist his desire for creation, he asked if they were content with living like that: wouldn't they be interested in other forms of life around? They were promptly tempted, and they set out about the activity of creation. The elder children they procreated, described in the account as of the elder sister, were all well to do. They included Brahmana, Sherry, Moily, Billava and others. But the younger children, of the 'younger sister', were poorer as they were the children of an incestuous union of brother and sister. Narayana ordered the younger ones to subsist on the leftover food from their elder siblings.[23]

Another version is more or less in the same pattern. During a ferocious flood a brother and a sister were floating on a log of wood. Life was very hard for them as there was nothing to live on, God appeared before them on the fourth day and asked them how they were related to one another. 'Man and wife', said the boy, 'please bless us'. God blessed them and disappeared. The process of creation began. All was green. The couple begot three hundred children. God was immensely pleased. He gathered them around him and ordered them into four divisions, assigning different jobs to each. Thus emerged Brahmanas, Kshatriyas, Vaisyas and Sudras. Four pairs of boys and girls were left out in the end. God decreed them to be Koragas and enjoined them to depend on their brothers for their livelihood.

There are interesting themes in all these accounts. To begin with, the insistence on origin from a brother and a sister is significant, because this is thought to be crucial for purposes of maintaining purity of descent.[24] In fact, this is typical of all 'primitive societies'. Secondly, these traditions state that those who claim positions in the upper castes in reality share the same origin as the less fortunate sections of society and that it was on account of a silly mistake on the part of god that society got divided on existing lines. These statements, to be sure, are not made to claim equality by challenging the positions of those in the upper classes; their purpose is to soothe the wounds of their degradation, something of a cool ointment. Those with nothing to be proud about in the present and nothing to hope for about future heave a sigh about the splendour of their past, more imagined than real.

They are not to be taken as actual statements of fact; going behind such expressions of wish-fulfilment, particularly in the ones in which they are represented as the original masters of the land but cheated out of their position by an act of treachery, one has to see elements of protest, veiled enough not to irritate the masters. The 'prose of counter-insurgency', employed in the Brahmanical versions presenting Hubbashika and his followers as so many erring villains who were promptly punished by adopting a suitable strategy, falls into the same pattern. At this point it is important to place the Habbashika story within the context in which it occurs in the Brahmanical account: that of the standard of revolt raised by the 'fallen Brahmanas' and that of the origin of these 'fallen Brahmanas'. It is really amazing that the Brahmanical documents contain these memories, whether received or constructed, which do not exactly do credit to them.

If one looks at it all as so many 'hidden transcripts' of resistance by the depressed, then one can read the myth of Koraga Taniya together with it. Koraga Taniya is a *bhuta* worshipped in South Canara. Its *paddana* gives an account of how the *bhuta* came into being. Taniya was a Koraga boy, orphaned at a very young age. One day, as he was playing alone, a Billava lady and her children came that way. Asked what he was doing there and which lineage he belonged to, the boy gave the name of a lineage, which happened to be her own. She took pity on him, took him home and brought him up into a strong and capable man. He would do the work of seven men. One day, when the feast in a *bhuta* shrine was going on, that family had to send provisions. Taniya carried seven head loads to the shrine all alone in one go. He was asked to leave them outside the shrine, but he insisted on carrying them inside. If he could carry them all the way to the gate, he could as well take them inside. When he reached the shrine, there was a tree full of citric fruits. He climbed it to pluck a few, for his foster mother was complaining the other day that they had run out of all pickles. As he climbed the tree, Maisandaya and Kinni Kodenganara, the two *daivas* presiding over the shrine, looked at him from the corner of their eyes and transformed him into a disembodied spirit. He became a *bhuta*, ever ready to help people out of distress, receiving offerings in the form of *bhuta kolas* and cooked rice mixed with turmeric powder.[25]

It is clear that a form of social protest is visible even here. The *daivas* punish Koraga Taniya for insisting on his right to take the offerings all the way to the shrine; even his body was not available any more! But, as he or his people could do nothing about it, he comes back to this world as a *bhuta,* his noble revenge being helping people in distress. It is significant that the costumes of the *bhuta* are simple and, that it is performed only on moonlit nights with no lamps, indicating how much of a 'subaltern' deity he is![26]

Attempts at identifying the Koragas as a caste have been going on from very early times. The Brahmanical accounts of the *Gramapaddhati* include them in the list of untouchable castes, created following the punishment meted out to Hubbashika and his men. Ramappa Vormika, the knowledgeable informant of Buchanan, thought of them in the late eighteenth century as belonging to the 'slave caste', of which there were fifteen subdivisions. The Kappada Koraga and the Soppina Koraga formed the two bottom-most in that list.[27] The characterization of the Koragas in the official documents has oscillated from tribe to caste ever since; but even when they were described as a 'tribe', they were presented as 'untouchables', 'a caste which pollutes without touching'.[28] The Brahmanical accounts mentioned above as well as the colonial writings, valorizing the Brahmanical discourse, as indeed they are ultimately derived from it, represent them as a caste, even though the expression 'tribe' is used occasionally in a loose manner, without any theoretical rigour. This has continued, even in the most recent accounts.[29]

The question, however, is: 'How can we really characterize them?' It may be useful to look at some of their features before classifying them. The Koragas are a matrilineal group, organized into several endogamous *balis* or lineages. There are divisions among the Koragas such as Ande Koraga, Kadu Koraga, Kappade Koraga, Soppu Koraga, etc. They make a living by working bamboo and making baskets. Some of them take up agricultural labour and a few others work as scavengers in the settlements of the 'upper castes', removing the carcass of animals from settled areas being one of the jobs they did. In recent years, some have taken to modern education and have found petty jobs in government services. They speak a separate dialect, which is very close to Tulu.[30] Traditionally they are worshippers of Sun, the naming of their male children, depending on the weekday of their birth. Thus, they give the names Aitha, Coma, Angara, Mudda, Gurva, Thukra, and Caniya respectively for those born on Sunday through Saturday. They too are, like the rest of them in the Tula country, worshippers of the *bhutas*. In recent years, however, they have started worshipping the typically 'Brahmanical' deities, and those that can afford make a rare pilgrimage to Sabarimala. A few have taken to the Gospel, whose condition, however, is not any better than when they were still 'heathens'. Eating habits, rituals on childbirth, marriage and death, etc., are somewhat exclusive; and so is the insistence on fidelity in the matter of sex.

All these, and many other features which can be had from the rich literature on Koragas, make them a people of considerable interest. Does it mean that they are a 'tribe'? It must be remembered that they are part of a highly stratified society in which caste is the marker of stratification. They are placed at a specified point, usually at the bottom of the scale, within the caste hierarchy where the point of reference in defining status is the relation with and proximity to Brahmanical groups. It is equally important that they are not only assigned this position in the hierarchy from above, as it were, but also that they accept it, as indeed their own mythology and current practices indicate. They have their defined relationship with other castes of varying ritual and social status. They, in fact, form part of a world where a pre-caste existence is not possible. Official labels such as 'primitive tribe' and officially sponsored projects for the welfare of the people are certainly welcome; but these official descriptions hardly help in understanding the reality. That the Koragas are living as a caste with all its features in a caste society does not mean anything about their economic and social position. If historians and other social scientists realize that 'tribe', 'caste', etc., are categories used for classifying peoples for the sake of understanding them, they will be able to accept that being the one or the other has not meant anything for those people. They continue to live in their own world, looking from the hardships of their gloomy present back to the imagined glory of a conjured up past and forward to the real horrors of an uncertain future.

NOTES

1. There is a rich literature on the Koragas. For instance, I list the following in chronological order:
   Francis Buchanan, *A Journey from Madras through Mysore, Canara and Malabar,* Vol. III, London, 1807; Edward Balfour, *The Cyclopaedia of India,* London, 1858; Ullal Raghavendra Rao, 'The Koragas', in *Indian Antiquary,* Vol. II, 1874, pp. 195-9; *Administration Reports of the Madras Presidency,* 1893; Gustav Oppert, *The Original Inhabitants ofBharata Varsha,* London, 1893; J. Sturrock, *South Canara,* Madras District Manuals, Madras, 1894, pp. 176-8; Edgar Thurston, *Castes and Tribes of Southern India,* Vol. III, Madras, 1909; B.A. Saletore, *Ancient Karnataka,* Vol. I: *History ofTuluva*, Poona, 1936; D.N.S. Bhat, *The Koraga Language,* Poona, 1971; Amrita Someswar, *Koragaru,* Bangalore, 1982;

Chandrashekhara Kambara, ed., *Kannada Janapada Vishwakosha*, Bangalore, 1985, pp. 519-23; K. Chinnappa Gowda, ed., *Paniyara*, Mangalore, 1989; idem., *Bhutaradhane*, Mangalagangotri, 1990; A.V.Navada, *Janapada Samalocane*, Bangalore, 1993; Purushothama Bilimale, *Koragara Samskrti*, Bangalore, 1993; P.Kamalaksha, *Dakshina Kannada Jilleya Harijana Mattu Girijanara Samajika Ithihasa*, Bangalore, 1994; Nagendra E. Rao, 'The Historical Tradition of South Canara and the Brahmanical Groups: A Study of Gramapaddhati and Sahyadri Khanda', M. Phil. Dissertation, Mangalore University, 1995; C. Vasudevan, *The Koragas—The Forgotten Lot: The Primitive Tribe ofTulu Nadu*, Trichur, 1998; Navin Kumar Marike, *Tulunadinalli Koraga Taniya*, Puthur, 2000; Lawrence D'Souza, ed., *Nammo Bhase* (Koraga language), Mangalore, n.d.; idem., *Koragas: A Primitive Tribe of South India*, Mangalore, n.d.; Mohammed Peer, 'Social, Economic and Educational Conditions of the Koragas: An Action Plan', Report on a project assigned by the Jilla Panchayath, Mangalore, n.d. Besides, there is a touching poem on the Koragas by Kayyara Kinhanna Rai.

2. Francis Buchanan, *A Journey from Madras through Mysore, Canara and Malabar*, Vol. III, London, 1807, pp. 100 ff.
3. Edward Balfour, *The Cyclopaedia of India*, London, 1858, p. 599, quoted in C. Vasudevan, *The Koragas - The Forgotten Lot: The Primitive Tribe of Tulu Nadu*, Trichur, 1998, p. 49.
4. M.T. Wallhouse, quoted in Edgar Thurston, *Castes and Tribes of Southern India*, Vol. III, Madras, 1909, p. 424.
5. Gustav Oppert, *The Original Inhabitants of Bharata Varsha*, London, 1893, pp. 168-80.
6. Quoted in Vasudevan, op. cit., Appendix VII, pp. 184-5.
7. *Administration Reports of the Madras Presidency*, 1893, p. 225, quoted in Vasudevan, op. cit., p. 33.
8. Thurston, op. cit., p. 424.
9. B.A. Saletore, *Ancient Karnataka*, Vol. I: *History of Tuluva*, Poona, 1936, pp. 52-3; idem, *Wild Tribes in Indian History*.
10. Purushothama Bilimale, *Koragara Samskrti*, Bangalore, 1993, pp. 9-15.
11. Ibid., p. 11.
12. Ibid., p. 16, *aithihya* 3.
13. Ibid. *aithihya* 4.
14. Ibid. *aithihya* 5.
15. Ibid. *aithihya* 6.
16. Ibid. *aithihya* 7.
17. Ibid. *aithihya* 8.
18. There are several versions of the *Gramapaddhati*. For an analytical study of *Gramapaddhati* as representing the historical tradition of the Brahmanical groups in South Canara, Nagendra E. Rao, 'The Historical Tradition of South Canara and the Brahmanical Groups: A Study of Gramapaddhati and Sahyadri Khanda', M. Phil. Dissertation, Mangalore University, 1995.
19. An account of this is available in all versions of the *Gramapaddhati*. I have used the edition, Krishanayya Holla, ed., *Gramapaddhati*, Bappanadu, 1924.
20. For a discussion of these dates, Vasudevan, op. cit., pp. 37-39. Vasudevan, however, takes the historicity of Hubbashika as sacrosanct.
21. Oppert cited in ibid, pp. 29-30.

22. ibid, pp. 30, 39-43. Vasudevan is a strong advocate of this 'identification'.
23. Bilimale, op. cit., p. 27, *aithihya* 1.
24. Whole descent groups being described as the progeny of a brother and a sister is usual. For a discussion, Romila Thapar, *Ancient Indian Social History: Some Interpretations,* Delhi, 1978, pp. 310-12, 314.
25. Someswar, op. cit.; pp. 38-47; Marike, op. cit.
26. K. Chinnappa Gowda, *Bhutaradhane,* Mangalagangotri, 1990, p. 141.
27. Cited in Vasudevan, op. cit., p. 18.
28. Cited in ibid, p. 33.
29. Mohammed Peer, op. cit.
30. D.N.S. Bhat, op. cit; Lawrence D'Souza, ed., *Nammo Bhase*, op. cit.

Tribes, Forest, and Social Formation in Early South India

Rajan Gurukkal

Occasions sometimes enable us to look at the old sources differently and gain new visibility quite worth the exercise. It is such a gratifying occasion to try and understand the place of tribes and forest in the early historic social formation of the Tamil macro region that is known in ancient Tamil heroic poems[1] as Tamilakam, geographically the landmass between Venkatam and Kanyakumari. Early historic, often called pre-Pallavan, refers to the period from the closing centuries of first millennium BC to about fifth century AD. In the paper the term tribes is intended to denote descent groups. The term forest is used in its generic sense meaning untamed land covered by trees and shrubs as natural growth. Theoretically a social formation refers to the interactive coexistence of two or more economies structured by the dominance of one. It is used here as an eminent conceptual framework for understanding the social history of any region where peoples of unevenly developed economies coexisted.

Any study of social formation at the outset necessitates knowledge about the material processes of human adaptation to the different ecosystems of the region and the social processes of appropriation. The identification of the peoples, their forms of subsistence/economies, technologies, kinds of labour and institutions of its realization and patterns of the distribution of resources and power becomes the starting point for the study. Along with these it necessitates further knowledge in the same line about questions such as the nature of interaction among the various economies and the institutional/structural means evolved through the process, for us to identify the dominant economy and understand how it dominated the rest.

Our knowledge about the economic and socio-political processes of the space and time under review is almost entirely based on allusions in the Tamil heroic literature. This is not altogether accidental because the rise of the social formation under study synchronizes with the emergence of the Tamil heroic literary tradition. Other corroborative sources like Brahmi label inscriptions, foreign notices and numismatic material jointly with heroic compositions push the social formation back to around second century BC while Iron Age archaeology assigns its early phase to the beginning of the first millennium BC.[2] Generally historians accept the turn of the Christian era as its central period.

We have some amount of clarity about the nature of human adaptation to the ecosystems of ancient *Tamilakam*, thanks to the *aitinai* concept of the Tamil heroic poetics. In the poetic concept, the Tamil region is a physiographical assemblage of five landscape eco-types, viz., *kuriñji* or the forested hilly area, *mullai* or the pastoral tracts, *pālai* or the parched area, *marutam* or the wet-rice fields and *neytal* or the littoral inhabited by hunters, shifting cultivators, pastoralists, cattle-lifters, agriculturists and fisherfolk and salt makers respectively. The concept helps us imagine the material processes of the ecologically determined human existence and coexistence possible in the region in ancient days. In the region the inhabitants were, as the heroic poems speak, the descent groups such as *kuṛavar*, *veṭṭuvar*, and *vēṭar* who subsisted on hunting/gathering and shifting cultivation; *iṭaiyar* who subsisted on agro-pastoralism; *maṛavar* who lived on predatory means; and *paratavar* who subsisted on fishing and salt manufacturing.[3] There were several groups of functionaries too such as *uḻavar* (wet-rice agriculturists), *toḻuvar* (cultivators of dry land called *puṅam* or *ēṅal*), *taccar* (carpenters), *kollar* (smiths), *vanikar* (traders) and a few others as the allusions in the heroic poems would have us believe. What emerges is an assemblage of coexistence and interaction of various tribes following the means of subsistence determined by the landscape ecosystems (*tiṇais*) that they inhabited.

The findings of the past researchers help us conceptualize the above scenario in terms of coexistence and interaction of multiple economies, peoples and the emergent power structure for characterizing it as a social formation.[4] This is a social formation with hunting/gathering/fishing, agro-pastoralism, wet-rice agriculture, salt manufacturing, crafts production and exchange as the constituent economies, of course not as mutually exclusive status, but of overlap.[5] The technology behind the economies was related to the multiple uses of iron. In all the economies productive and distributive relations were based on kinship. The principal industries of the social formation were cattle rearing, the cultivation of millet, ragi and paddy and, the production of a variety of crafts. Though paddy cultivation was the most superior industry of the times both in terms of productivity and technology, it was yet to articulate the conditions for its dominance. Its relations of production, stuck to kinship were incompatible to development. The dominant economy of the social formation was agro-pastoralism that combined cattle keeping with shifting cultivation as borne out by the preponderance of references to cattle raids and associated practices, rites and rituals and the centrality of institutions like gift and redistribution besides the hegemony of heroic culture. Studies show that everything in the socio-cultural regime revolved round this economy and everything was disposed of so as to establish its hegemony over other economies.[6]

Organized under spontaneously evolved self-sustaining kin units (*urs*) of production, the tribes maintained exchange relations (primarily of the goods for goods type) and shared-cultural practices, which eventually led to the making of *Tamiḻakam* as a region of linguistic homogeneity. Exchange of goods and services hardly involved the concept of exchange-value/price and the peoples seem to have had no notion of interest/profit. It is true that both transmarine and inland traders from far away places like Rome and Gangetic valley had entered into exchange relations with the region. Recent studies have shown that the long distance exchanges were also based on the principle of use-value rather than exchange-value.[7] Though Roman currency was extensively used in the exchanges, but only as part of contemporary valuables rather than as means of payment and measure of value. Roman money entered the transactions as goods of treasure that could be used for enhancing status and ranking but not as a productive facilitator of labour.

An aspect to be significantly singled out in the context of the present topic of discussion is that the landscape ecotypes of the dominant economy were all within the forest ecosystem. The pastoral tracts (*mullaittiṇai*) and tracts of shifting cultivation (*kuṟiñjittiṇai*) that jointly constituted the larger terrain (*vanpulam*) of the economy were grassy hill tops and vegetated slopes of the forests. The parched zone (*pālaittiṇai*) that enlarged in summer and dwindled during monsoon and inhabited by predatory clans called *maravar*, was also a part of the forest. In short, excepting the terrain of wet-rice agriculture, primarily confined to small pockets of fertile river planes (*mēnpulam*) and the littoral, most of contemporary human habitats were in the forest. Several allusions in the Tamil heroic poems show that contemporary human settlements and forests existed as subsumed by each other. The *aintiṇai* concept differentiating the terrain in terms of five eco-types (*tiṇai-s*) involves no rigid contrast between forest and non-forest. It is clear that peoples seldom conceived forestland (*kāṭu*) in contradistinction to human inhabited land (*nāṭu*). The *katu* >< *nāṭu* opposition as it existed in the landscape consciousness of the later agriculturists was not known to them.

The poems clearly show that the political level of the social formation was characterized by tribal chiefdoms with a pattern of distribution of power varying from the simple to the complex along the small and big descent groups.[9] The *kuṭimākkal* or domestic segments of a descent group and their *ko-mān* or *perumakan* (chief) constituted the simplest structure that signified an organized settlement or *ur* bound by kinship. The basic constituent in the structure was *kuti* or family. The heroic poems unveil before us an active scenario of coexistence and interaction of unevenly evolved chiefdoms of three types, viz., the *Kiḻār, Vēḷir* and *Vēntar*. Most of the *Vēḷir* and *Kiḻār* were chieftains of the forest. One of the *Vēntar-s*, namely the Ceras were chieftains of forest lands too.

Several poems indicate that it was in the pastoral tracts (*mullai*) and hills (*kuṛiñji*) in the forest that the *Vēḷir* chieftains held sway over. The poems address them as hill chieftains (*malaiyamān*) heading the descent groups called *vetar, itaiyar* and *kuravar. Venkaṭamalai, Kanṭiramalai*, Kollimalai, Mutiramalai, Kutiramalai, *Paṛampumalai*, Potiyilmalai, *Pāyarmalai, Eḻilmalai* and *Nñjilmalai* are the famous millet rich hill chiefdoms celebrated in the poems (*PN*: 143, 168). *Eḻilmalai* was the most prominent hill chiefdom of Kerala and the lineage of Nannan, the hunter chief of *vēṭar* (*vētarkōmān*) was related to that of the chiefs of Kantiramalai. Another chiefdom closely linked to the southern end of Kerala was Potiyilmalai. A hill chief called Irunko-*vel*, one of the traditional five *vēḷs*, is mentioned in a poem as *vēṭarkōmān*, the chief of *vēṭar*, to have belonged to a long line of 49 generations of chiefs (*PN*: 202, 201). The poems celebrate the *Āy* as ***kuṛavarperumakan***, the chief of ***kuṛavar*** in the hill called Potiyilmalai rich in honey, jack fruit, elephants and monkey (*P.N*:127-36). The *Āy* chief is addressed as *māvēḷ*, the big *vēḷ* and mentioned as belonging to the *Āykuti* (the *Āy* family). The association of the term '*Āy*' with *āyar* (pastoralists) and the claim of the later *Āy* chiefs to have belonged to the *vriṣhṇikula* are there but as such there is no direct evidence to show that they were pastoral chiefs. Pari, the chief of Parampumalai; *Ōri*, the chief of Kollimalai; *Kāri* who killed *Ōri* and became the chief of his hill, *Eḻini*, the chief of Kutirmalai and *Pēkan*, the chief of Vanmalai, *Kumaṇan*, and the chief of Mutiramalai are the most celebrated hunter chiefs of *Vēṭar* or *Kuṛavar* (*P.N*: 158). Sometimes the hill chiefs are called *vēṭṭuvar*. This would suggest that the term *vel* derives from *vet*, meaning hunter.

All these chiefs, uneven amongst themselves as regards their possession of resources, had to resort to plunder raids. Most of them seem to have maintained predatory control over the agrarian zones in their proximity (*P.N*: 110, 168). The hill chiefs had plundered themselves against one another for accumulating resources for redistribution. Through the exchange of forest goods some of the hill chiefs could have procured new resources like prestige goods. Irunko-*vel's* hill is praised in a poem as gold yielding (*P.N*: 202) obviously meaning the exchange of its resources like ivory, monkey, animal skin, sandal wood, etc., for gold coins from Rome. But how exactly the chiefs exchanged the forest-goods and realized the returns is not clear. Nevertheless, the hills of certain chiefs had a relatively better potential of new resources, which is referred to in the poems with the term *yāṇar*. *Paṛampumalai* is called '*yāṇaraṛa aviyan malai*', the hill with the potential new resources (*P.N*:116). It is evident that the resource potential varied from chief to chief and obviously some of them were comparatively more resourceful. For instance, the *Āy* chief is said to have possessed horses and chariots. His house was called *Koil* and its surroundings as *nakar*. Pari is also mentioned to have possessed horses and chariots. *Piṭṭan Koṛran* of Mutiramalai is called '*kaṭumān koṛran*', the possessor of horse. Horse was not a common possession of chiefs and references to their giving of horses and chariot, as gifts seem to be a mere conventional mode of praising. Possession of rare goods always enriched the status and ranking of chiefs and in the case of the *Āy* who is known after his family name, such a higher status is explicit. No other hill chief is known after his family, except perhaps the *Cēras*, who seem to have emerged from the level of a hill chief.

The structure of the political level of the hill chiefs was essentially a simple one based on kinship. A chief is referred to in the poems as *kō-mān (kō-makan)* or *perumān (peru-makan)* of a given group of people as evidenced by *kuṛavar kōmān* or *kuṛavar-perumakan*, etc. Normally a hill comprised several settlements (*ur*) of the chief's people and a few others the size of which varied depending on the extent of the hill and the form of subsistence. These others included the *kuṭis* of magico-religious and other functionaries like *tuṭiyar* who made a kind of drum called *tuṭi* and *panar* who composed *pāṭṭu* or songs (*PN*: 269, 280, 285, 291). The range of redistributive relationship of a hill chief was limited, though a few bards from distant places also met him occasionally. Similarly, his predatory range was also small and could have afforded only small-scale raids. With no relations transcending kinship, involved in the form of production and resource circulation, the political power of this category of chiefship remained subsumed within the kinship system.

Like the *Vēḷir* chiefs, the *kiḻār* chiefs were also hunter chiefs either of *veṭar* or *kuṛavar* tribes. The poems mention them as the *ūr-kiḻār* or *ūr-maññar* who were generally chiefs of small settlements, mostly in *vanpulam*, the forest land (*PN*: 177, 180, 181). The *ūr-kiḻān* of a pristine type must have been a smaller variant of the *malai-kiḻavōn* or the hill chief. A poem praises the *kiḻān* of *Īrntūr*, a settlement of *vanpulam* with marginal resources, depending primarily on plunder raids, as the enemy of hunger who would summon his blacksmith on seeing a hungry bard and order for a new lance to go for a raid to appease the bard's hunger (*PN*: 180). Certain *kiḻār* are also mentioned to have held sway over agrarian tracts who were relatively more resourceful (*PN*: 176, 376, 381, 388). However, they also had to maintain predatory control over other settlements to meet the redistributive needs. A few *kiḻār* seem to have functioned like bards of the well-to-do chiefs. Most of the *ūr-kiḻār* and *ūr-maññar* were subordinates of bigger chiefs and had the obligation of fighting for them.

Of the three *Vēntar* variety of chieftains represented by the chiefly lineages, viz., *Cēra*, *Pāndya* and *Cōḻa*, the Ceras are referred to in the poems as '*kāñaka-ñāṭan*' (the chief of the forested *natu*) or *malaiyan* (the chief of *malai* or hill), which is suggestive of their ecological region. A poet praising *Cēramān Kōtai Mārpan*, expresses confusion as to how the chief should really be addressed (*PN*: 49). The poet asks whether the chief could be called *nāṭan* as he had *marutam* lands or *uran* as he had *kuṟiñji* lands or *cērpan* as he had coastal tracts. This would suggest that the *Cēra* region was a mixture of diverse ecological zones with the predominance of hills and forests. The resource base of the *Cēra* was also therefore diverse though forest wealth was the main. A poem incidentally refers to the hill products (*malaittāram*) and sea products (*kaṭ-aṟṟāram*) of *Cēran Cenkuṭṭuvan* and the gold that reached ashore by boats (*PN*: 343). The *Panḍya* also had a mixed ecological region dominated by pastoral and coastal tracts. A *Panḍya* chieftain calls himself the head of the land of numerous new resources, '*yanar maiyar koman*' (*PN*: 71). The *Cōla* who is well known as '*kāviri kiḻavōn*' in the poems had his land in the Kaveri delta, rich in paddy and sugarcane (*PN*: 61).

The poems address *Vēntar*as *kāvalar*(protector) of the *kuṭimākkal*, the settlers. *Pānḍya Neṭunceḻiyan* refers to his *kuṭimākkal* as '*en niḻal vāḻnar*', meaning those living under his shade. This would presuppose the exaction of something in return from the *kuṭimākkal* for the protection offered to them. In the case of the Cola it is clear that the *Vēntan* used to exact *puravu* (paddy) from the *kuṭimākkal* (*PN*: 75). All the three *Vēntar*are referred to in the poems with the term *iṟaivan* which means he who exacts. This would suggest that they had exacted from the people what was feasible according to the resource potential of the region. However, it appears that even this kind of exaction had a predatory character. There is no evidence for any regular periodic exaction in fixed measure or quantity by any of these chieftains. So we may safely assume that the *Vēntar* exacted resources through predatory operations and as prestations and voluntary offerings. Poems belonging to the *ceviyaṟivuṟu* and *porunmoḻikkāñci tuṟais* advise the chieftains how to keep the settlements productive and how to appropriate their surplus in a sustainable manner. A poem advises a *Vēntan* not to behave like an elephant in the sugar cane field that destroys much more than what it eats. These are obviously exhortations by poets who have known the instituted modes of periodic exaction in developed kingdoms beyond Tamilakam.

That the *Vēntar* category of chieftains also appropriated the resources through predatory means and voluntary offerings like prestations, shows that they were not far removed from the milieu of hill chieftains. There are indications in the poems to the expansion of these powers from their original *ūr (mutūr)* obviously through the process of subjugation (*P.N*: 54). The poet shows Ceraman Kuttuvan Kota, sitting as the *uṭaiyor*(lord) of a *mutūr* in the place of its original chief, probably suggesting subjugation. The subjugation process seems to have involved three different methods, viz., subordination with tributary obligations, expulsion and marital alliances. There are many references in the poems to all these methods of enlarging the domain of the *Vēntar. Vaḷḷuvan*, the chief of *Nānjilmalai*, is mentioned in a poem as a *Cēra* subordinate with military obligations (*P.N*: 139). The chiefs of *Pāyar* malai and *Veṭṭaru* were the other known *Cēra* subordinates. Similarly *Nākan*, the *Kiḻavan* of Nalai and Nampi *Neṭunceḻiyan* are mentioned as *Pāṇḍya* subordinates with military obligations (*P.N*: 179, 239). There are a few poems in praise of the chiefs *ēnāti Tirukkuṭṭuvan, ēnāti* Tirukkilli, and *enati* Tirukkannan as *ēnāti* of the *Cōḻa* (*P.N*: 167, 174, 394). *Paṇṇam*, the *kiḻan* of *Ciṛukuṭi* and Aruvantai the *kiḻān* of Ampar were the *Cōḻa* subordinates with tributary and military obligations. Sometimes chiefs in the fringes were subordinated by two *Vēntar* and naturally this subordination fluctuated from one to the other (*P.N*: 380). In addition to such important subordinates, there were numerous minor chiefs called *Cīṟūr-mannar*, mostly *maṛava* headmen of forest hamlets, who functioned as the miliatary chiefs of the *Vēntar*, particularly the *Pāṇḍya* and *Cēra*.

The returns from exchange relations must have enabled the *Vēntar* to possess gold and other prestige items. As already noted, it is not clear how they were involved in the process of exchange. The poems show that the major activity of the *Vēntar* like the *Vēḻir* was accumulation of resources and their redistribution following the determinate pattern of social relationships. Plunder was indispensable for them also since their redistributive network was much more elaborate and complex than what they could have afforded with their actual resources. They had a large body of dependants such as their kinsmen (*kiḻaiñar*), scholarly bards, (*pulavar*), warrior chiefs (*maṛavar, kiḻār* and *mannar*), warrior men (*maṛavar*), bards (*pāṇar* and *poruṇar*) magico-religious functionaries and so on. The poetic flower symbolism of *vetci* (cattle raid), *karantai* (cattle recovery), *vañji* (chieftain's raid), *kañji* (chieftain's resistance of a raid), and *tumpai* (preparation for raid) show how institutionalized and common the plunder was.

The structure of the *Vēntar* level of political power was relatively more complex since its redistributive social relationship was elaborate. It involved some kind of a simple hierarchy from *Vēntar* to the *kuṭimākkal* with *kiḻār* or *mannar* intermediaries. The hierarchy cut across kinship and distanced the *Vēntar* from *kuṭimākkal*. But they did drink and dine with the *maṛavar* during *uṇṭāṭṭu*, the pre-raid or post-raid feasting at the residences of *Vēntar*. A set up of complex redistributive political economy based on raids, precludes the formation of a structured polity with defined positions and functions.

There seems to be a lot of difference between the image that the poems try to secure for the *Vēntar* and reality about them. We know that unlike each of them claimed, the whole of *Tamiḻakam* did not belong to them and there were other tribute-receiving hill chiefs like Atiyaman who were almost nearer to *Vēntar* in status. A poem in praise of *Netumān* Anji warns all the chiefs of agrarian settlements to rush to him *tiṛai* if they wished to keep their *ur* with them (*PN*: 97). Many of the hill chiefs were uncompromisingly opposed to the *Vēntar*. *Pāri* of *Paṛampumalai* is one good example. He offered strong resistance to the *Vēntar*, though he was subsequently defeated and killed. So reality was that *Vēntar* were also chiefs, but of a little higher category. The crucial difference was their relatively greater resource power, larger redistributive social relationship and better sources of legitimation. They were surrounded by Brahmin *pulavar* of the Vedic tradition and a few were well informed of the *śāstrāic*, *purānic* notions of kingship. But a predatory chieftain, whose status and power were linked up with the range of redistributive social network, could have hardly gone by *śāstrāic* prescriptions. A poem by a scholarly bard reminds Pandya Nanmaran of the fact that '*aṛa neri mutaṛṛē aracin koṛṛam*' meaning greatness of royalty remained with the primacy of '*dharma*' (*PN*: 55). All songs in *Ceviyaṛivuṛu tuṛai* contain ideas of this type, which sounded exotic in a milieu of plunder raids and redistribution. The *Vēntar* drew upon heroism and gift-giving for ideological force. Bards were the main strength behind their name, fame and legitimacy. They kept the image of the *Vēntar* by roaming round the land with their songs in praise of the latter's exploits. The *paṇāṟṟuppaṭai* category of poems itself exemplifies the instituted nature of such circuits.

Largely the headship over clan kin ties, i.e. the chiefly power, constituted the source of authority to command collective labour.

Chieftains of higher levels could command the labour of their mercenaries (*viṭutoḷil*) who seem to have included potters and metalworkers. Probably the most subjected kind of labour they commanded must be of the captives. Possessing warrior power, they could probably command all kinds of specialized labour. There are references in the poems to warriors (*maṛavar*) offering paid service of protection to caravan troops of *umaṇac-cāttu* or salt merchants (*AN*: 39:10, 167:7, 245:6, and 291:15). In these poems the means of payment is mentioned as *dravya* (wealth) the exact connotation of which is unclear.

The principal social mode of labour realization was familial/cooperative. A few crafts like metal working and pottery that come under the category of skilled labour, must have been full-time trades of specialists and hence hereditary. Iron, the most extensively used metal, had a central place particularly as the base of weapons whose significance in a predatory society is explicit. Moreover, the practice of burying iron objects along with the dead had pushed a great deal of iron out of circulation presupposing continuous iron working as a full-time occupation of hereditary specialization. The production of earthen pots, a characteristically brittle artefact, was obviously a continuous full-time activity, for their use was extensive both for the living as well as the dead. The number of such full-time artisans/crafts folk of hereditary occupations was relatively more in the headquarters of bigger chieftains of the *vēḷir* and *vēntar* levels. As the major redistributive pools of resources, the chiefly settlements could support more full-time crafts. Another full-time function of hereditary nature was that of warriors (*maṛavar*). Every settlement (*ūr*) needed full-time warriors since the main mode of political appropriation of resources was predatory. In association with the chiefly households, three other full-time hereditary functionaries, viz., the *pāṇar* (bards), *paṛaiyar* (who play a kind of raid drum called *paṛa*) and *tuṭiyar* (who play a small drum called *tuṭi*) were there.

The preceding discussions of the features of the economic and political levels of the social formation clearly indicate that they were predominantly characterized by the tribal polity and forest cultures. This is further endorsed by the various rites and rituals and instituted practices of the times. So do the various beliefs and customs, costumes and ornaments that characterized the regime of culture. Institutions for the chiefly celebration through group drinking and dining of warriors after and before cattle raids (*unṭāṭṭu*), chieftains' public rice-feasts (*cōṛṛuviḻāvu-s*) and female dance propitiating the battle goddess (*kuravaikkūttu*) are some of the examples to mention only a few. The primacy of forest culture and tribal way of life continued until the social formation got dissolved itself into a new one dominated by advanced wet-rice agriculture.

The dissolution of the tribes and forest social formation began to manifest itself in the form of a series of institutional and structural changes. Predatory marches of chiefs, their ravaging of settlements, redistribution of *ūr,* the consequent migration and subsequent immigration and so on, were some of the ongoing events of transforming effect on the social formation that was generating various contradictions within due to built-in factors like kin labour and redistribution. The most striking contradiction was the continued articulation of conditions totally uncongenial to the development of plough agriculture, which was the most potential form among contemporary forms of production. Predatory marches of chieftains, their destruction of agrarian settlements as part of the scorched earth policy in raids, and the dominance of the ideology of war and booty redistribution provided an adverse circumstance for the development of agriculture. As we have already seen, redistribution exerted pressure on production, but failed to translate itself as a motor of intensified production since there was no scope for it within the kinship based forms of production. Intensified labour mobilization for better production was beyond the working power of contemporary political apparatus that had little coercive ability. It was not possible for the social formation to persist on for a long time in a set up of complex redistribution, generating contradictions. Obviously the major trend in the process was that of the gradual dissolution of the social formation. The process involved the expansion of wet-rice fields, slow disintegration of tribes into domestic segments of hereditary arts and crafts, formation of agrarian villages as clusters of settlements (*cēri-s*), occupied by artisans, craftsmen and tillers, and owned and controlled by landed households. A corresponding manifestation was the receding of the forest from the mainstream life-world and the rise of agrarian landscape as its contrast.

The seeds of the dissolution of the tribes and forest social formation can be discerned in the redistribution of resources particularly land beyond the social relationship of kinship, which had certain lasting consequences. It appears that at some point of time the institution of redistribution involved gift of land, mainly to warrior chiefs. In the case of the warrior chiefs the gift must have meant only transfer of predatory control rather than ownership. As scholarly Brahmanas were part of the redistributive social relationship, land seems to have been given as gift to them too, though not extensively. Being not cultivators by themselves, Brahmanas had to get their land worked by others. This implied the making of a new system of relations in production, transcending the framework of kinship. We have a few references to prominent Brahman households. Kauniyan Vinnan Tayan of *Puñjāṟṟūr*, is an example. At the same time he appears in the poems as a householder with pastoral wealth rather than agrarian resource (*PN*: 166). It is a fact that with their system of non-kin relations of labour the Brahmana households witnessed the beginnings of social stratification.

What began taking shape in the Brahman households was crucial for the real beginnings of a hierarchy. The permanent workforce attached to the Brahman households had the greatest possibility of being conceived hierarchically because of the stratifying system of production relations and the Brahmanical tradition of social differentiation. The notion of hierarchy was implicit in the system of production in which the relation between two objectively antagonistic classes was fundamental. The mid-first millennium AD was thus a turning point in terms of stratification and hierarchical ordering. The process took more than two centuries to characterize the social aggregate. During the fifth-sixth centuries the agrarian societies of Tamilakam were perceptibly becoming class-structured. This was directly related to the spread of plough agriculture and the corresponding new relations of production that meant social stratification based on entitlements to the nature of land-use. It was primarily a tripartite stratification of the people into landholders, leaseholders and tillers. Since plough agriculture also meant specialization of a variety of arts and crafts, a further stratification of the people who were grouped along the line of occupations followed gradually. The nature of rights over land and the level of entitlement to the produce determined the strata of the people of different arts and crafts. Divided largely into the upper and lower strata the people were soon woven into a system of differentiation within each stratum. The social relations of the period resulted from an aggregate of these.

The social relations began to be further structured during the sixth-seventh centuries with the steady expansion of plough agriculture across the wetland. Expansion of agrarian settlements through the creation of *brahmadēyas* often involved the superimposition of the superior rights of the Brahmanas over the communal holdings and the clan families of the locality. It must have been an intricate process of transformation of primitive agriculture and clan settlements into advanced agriculture and farmer settlements, respectively. The main features of the process were differentiation, stratification, and political formation leading to the development of the state-system and authority structures. These were simultaneous developments taking place as supplementary and complimentary to one another, resulting from the growth of paddy economy. Such developments were in their turn ensuring the further growth of the economy.

The relations of production in plough agriculture were expanding towards domination of the total society. This was a long institutional process involving the proliferation of occupational specialization and its ordering into a hierarchy.[10] The formation of agrarian localities was an ongoing process, and everywhere it accomplished a uniform structure of social relations. As agrarian expansion advanced, human settlements (*ūr*) originally bound by kinship got penetrated by the mechanisms of stratification. In short the transformation of non-Brahman villages into productive relations transcending kinship was a continuous process. The non-Brahman villages were called *vēḷānvakai* in contemporary inscriptions. Such settlements began to be integrated as agrarian localities (*nāṭu*). This *nāṭu* was hence fundamentally different from the *nāṭu* that figures in the heroic poems. As agrarian localities of hierarchically structured social relations, the *nāṭus* subsequently acquired great political significance in the monarchical system. It represented a clear contrast to the forest.

The disintegration of clan-kin ties became almost total along the agrarian tracts as paddy cultivation expanded with the corresponding developments like social stratification and widening of the division of labour through specialization.[11] In the process the clan identity disappeared and *jāti* came in as the substitute, mostly by retaining the names of the clan. Several names like *pāṇa, paṛaya, vēṭṭuva, vēṭa, kuṛava, maṛava, paratava*, etc., vouch for the retention of clan names for *jāti*. The first notable development towards the transformation of clans into *jātis* was specialization of labour. It was the practice of rewarding specialized labour with land-based entitlements that made specialization of labour hereditary. The reason was obviously the labourers' urge to perpetuate the entitlements. But how the labouring clans of hereditary specialization got absorbed into the *jāti* system cannot be explained easily. We know that the proliferation of hereditary occupations and their absorption into the *jāti* hierarchy were advancing side by side with the socio-spatial expansion of paddy cultivation. Also we know that at a later period new institutional agencies like the temple and service tenements were instrumental in crystallizing the *jāti* system. Further, the relation of the *jāti* system to Brahmanism is taken for granted by all. But at the same time the actual social process of the making of the *jāti* system is largely unclear and hardly made out as yet.

The society became class-structured and the dissolution of tribal chiefdoms began anticipating formation of the state with the new agrarian system articulating its polity. The new political formation represented by the Simhavarman line of the Pallavas and the Kdungon line of the *Pāṇḍyas* owed itself to the developing agrarian society whose expansion was linked to royal patronage. The *Cōḻas* of the Vijayalaya line at a later period represented the same kind of political authority engendered by the paddy-based economy. Though it is not clear whether the Ceras represented a comparable royal line of inheritance, the political authority represented by them too was engendered by the paddy economy. A perceptible institutional feature of agrarian expansion was the proliferation of *brahmadēya* villages throughout the fertile tracts of major river valleys in the region. This was an organized affair under the royal initiative. A few copper plates, say *Velvikkuṭi* plates for instance, speak about the restoration of the villages originally gifted to Brahmanas as *ēkabhōga* and subsequently lost through misappropriation by others. All such lost villages were later restored as *brahmadēyas* under the corporate control as the cases like the *Velvikkuṭi* vouch for. This shift from individual holding (*ēkabhōga*) to collective holding (corporate *brahmadēya*) is important in the context of the insecurity of the former. The proliferation of the latter meant the successful development of the new system of productive relations under a new institutional form and political patronage.

With the expansion of the new relations of production and the spread of wet-rice agriculture that became characteristic to the period from sixth-seventh centuries, the social formation structured by the dominance of forest economies came to an end. In short, the disappearance of the social formation dominated by forest economies involved a series of transitions like the ones transition from kin-labour to non-kin labour, multiple functionaries to hereditary occupation groups, clans to castes, simple clannish settlements to structured agrarian villages, and chiefdom to monarchy.[12]

NOTES

Abreviations: *PN* = *Puṛanāṉūṟu*, *AN* = *Akanāṉūṟu*, *Prp* = *Patiṟṟuppattu*.

1. 'Tamil Heroic Poems' as an expression refers to what is popularly known as the corpus of Sangam poems here. The corpus includes in its most archaic stratum some of the anthologies grouped under *Eṭṭuttokai* and *Pattuppāṭṭu* roughly belonging to second century BC and third century AD. For a detailed analysis of the structure, diction and context of the corpus, see K. Kailasapathy, *Tamil Heroic Poetry* (London, 1968).
2. Label inscriptions consist of the Tamil brahmi labels belonging to *c.* third century BC to fourth century AD. See I. Mahadevan, 'Corpus of Tamil Brahmi Inscriptions', in R. Nagaswamy (ed.), *Seminar on Inscriptions* (Madras, 1970). The foreign notices comprise mainly the Graeco-Roman writings. Archaeology of Iron Age in south India is a well-studied topic. See Guru Raja Rao, *The Megalithic Culture of South India* (Mysore: Prasaranga, 1972). L.S. Leshnik, *South Indian 'Megalithic' Burials: The Pandukal Complex* (Wiesbaden, 1974). Also A. Sundara, *The Early Chamber Tombs of South India* (Delhi, 1975). For studies in the cultural overlap and continuity between the periods of Megalithism and the early historic heroic society see, K.R. Srinivasan, 'The Megalithic Burial and Cairn Fields of South India in the Light of Tamil Literature and Tradition', *Ancient India*, No. 2, 1947, pp. 9-16. R. Champakalakshmi, 'Archaeology and Tamil Literary Tradition', *Puratatva*, No. 8, 1975-6, pp. 110-22.
3. See discussion of the various related aspects in K. Siva Thampy, 'Early South Indian Economy: The *Tinai* Concept', *Social Scientist*, 29, 1974, pp. 20-37. Also Rajan Gurukkal, 'Forms of Production and Forces of Change: Ancient Tamil Society', *Studies in History*, n.s., Vol. 5, No. 2, 1989, pp. 159-75.
4. For details see Rajan Gurukkal, 'Forms of Production and Forces of Change . . .', op. cit., pp. 160-8.
5. This is argued at length in Rajan Gurukkal, 'Forms of Production and Forces of Change . . .', op. cit., pp. 168 ff. For an analysis of the larger context see, Romila Thapar, 'Black Gold: South Asia and the Roman Maritime Trade', *South Asia*, n.s., XV, 2, pp. 1-28.
6. For a detailed consideration of the issue see Rajan Gurukkal, 'Forms of Production and Forces of Change . . .', op. cit., pp. 162 ff. Also his, 'Characterising Ancient Society: The Case of South India', Presidential Address, Indian History Congress, Ancient India Section, Patiala, 1998, pp. 24-5.

7. Ibid., pp. 21-3. For details of this process in the context of the transformations in Tamilakam see, Rajan Gurukkal, 'Forms of Production and Forces of Change . . .', op. cit.
8. See Rajan Gurukkal, 'Characterising Ancient Society: The Case of South India', Presidential Address, Indian History Congress, Ancient India Section, Patiala, 1998, pp. 24-5.
9. This is considered at length in Rajan Gurukkal, 'Antecedents of State Formation in South India', in Champakalakshmi, K. Veluthat and T.R. Venugopalan (eds.), *State and Society in Pre-modern South India* (Trissur, 2002), pp. 39 ff.
10. For a detailed consideration of the issue see Rajan Gurukkal, 'From Clan and Lineage to Hereditary Occupations and Caste in South India', in Dev Nathan (ed.), *From Tribe to Caste* (Shimla, 1997), pp. 205-22. Also, his 'Characterising Ancient Society: The Case of South India', Presidential Address, Indian History Congress, Ancient India Section, Patiala, 1998, pp. 24-5.
11. See Rajan Gurukkal, 'From Clan and Lineage to Hereditary Occupations . . .', op. cit., pp. 205-22.
12. Ibid., p. 220. Also see Rajan Gurukkal and Raghava Varier (eds.), *Cultural History of Kerala*, Vol. I (Trivandrum, 1999), pp. 257-63.

Situating Tribals in the Early History of Orissa

Shishir Kumar Panda

Orissa is a land inhabited by tribals who constitute 23.11 per cent of the state population. They are mostly found in almost every part of Orissa except the coastal region. The hill tracts of Orissa were the abode of Indian's original tribal population who continue to maintain their primordial cultural traits up to the present time. They have played a vital role in shaping the history and culture of Orissa. In due course of time the tribals have been assimilated into the Hindu society. There was a cross-cultural assimilation between the tribals and the Hindus which gave a new dimension to Orissan culture. Hence, the history and culture of Orissa cannot be assessed without examining the role of the tribals.

Though the development of history and culture of Orissa is intimately associated with the tribals, so far no major attempt has been made by scholars to unravel the 'tribal factor' in its history. In this paper an attempt has been made to identify the role of the tribals in the early history of Orissa. How did the tribals in due course of time assimilate into the Hindu society? Also, how the tribals chiefs of Orissa claimed the Kshatriya status and ruled over various kingdoms dynastically? Further, we shall examine their role in the socio-economic life of Orissa in the context of large-scale land grants to the Brahmanas in the tribal hinterland of Orissa and their upward mobility to the status of the Kshatriyas. The cross-cultural transmission between the tribals and the 'Hindus' will be dealt with in the context of the development of the cult of Jagannatha and the Shakti worship.

II

It is an accepted fact that in early times regions were named after different stocks of people. Ancient Indian literature, Brahmanical, Buddhist and foreign accounts while referring to the territories reiterate the people rather than the land. So the territories like Utkala, Kalinga, Odra and Kosala which roughly constituted modern Orissa had acquired their names after the tribal or the racial names of the people who inhabited them. Both the Buddhist *Suttas*[1] and *Pitakas*[2] speak of the Okkala or Utkala tribe whereas the epics and the Puranas[3] mention them as Utkala people. The *Mahavagga* of the *Vinaya Pitaka*[4] refers to the Ukkala Janapada, the homeland of Tapassu and Bhaloika, the first lay worshippers of lord Buddha whereas the *Mahachattarisaka Sutta*[5] of the *Majjima Nikaya* mentions that Vassa and Bhana were two tribes of Utkala people. Kalinga and Odra were also named after the tribes inhabiting these territories.[6] Pliny[7] mentions the three division of Kalinga tribes, such as Gangraides Calingae, Macco-Calingae and Calingae. Futhrer, he has associated Mount Meleus with the people called Monedes and Suari. A. Cunningham[8] has identified Monedes and Suari with the tribals Mundas and Shavaras who are even today inhabiting the upland regions of Orissa. Shavaras are too mentioned by Ptolemy as Sabarai.[9]

Among the tribes of Orissa, the Pulindas and the Shavaras were the most ancient. The Pulindas are mentioned along with the Shavaras in the *Aitareya Brahmana.*[10] It has been referred to in the *Mahabharata* that Sahadeva proceeded southward after defeating the Pulindas.[11] The *Charjya*[12] poems of the Oriya Vajrayana Siddhas composed between the ninth and thirteenth centuries mention a number of tribals such as Shavaras, Bhils and Dombs. Sarala Das also in his Oriya *Mahabharata*[13] described in details the life of the Shavaras and the Khonds. The *Jagamohana Ramayana*[14] of Balaram Das makes mention of Guhaka Shavara severally. Thus, literary evidence very much testifies to the habitation of tribals such as Shavaras, Pulindas, Gonds, Mundas, Bhuiyans, Khonds, Bhils, etc., in early Orissa.

Besides the literary evidences, the inscription also refers to the tribals of Orissa. The earliest inscription, the Ashokan edict[15] mentions the unconquered Aatavikas, the tribals of the Kalingan region. The Allahabad Pillar inscription[16] of Samudra Gupta referred to the forest region of Kosala and Mahakantara and its ruler Vyaghraraja. Shavara settlements too have been mentioned in the Narasinghpurcharter[17] of Samavamshi King Udyaotakesari Mahabhavagupta.

Further, from the inscriptional sources it is known that in early times several principalities grew in the tribal areas of Orissa. They were Mahendragiri,[18] Aatavika rajya,[19] Vidhyaatavi,[20] Mahakantara,[21] Kosala[22] and Jhadakhanda desha.[23] There were eighteen forest kingdoms (*ashtadasha-atavirajyam*) over which king Lokavigraha was ruling as known from the Kanasa copper plate grant.[24] Also, eighteen Gond principalities have been mentioned in the inscriptions[25] over which the Tungas were ruling with the epithet lord of the eighteen Gondramas (*ashtadasha-Gondra-Madhipati*). Perhaps, these principalities, later on were known as eighteen Garhjats in Orissa.

III

The most important development in early medieval Orissa was that some of the ruling families originating from the tribals later claimed the status of the Kshatriyas as known from their inscriptions. The Shailodbhava dynasty who ruled over Kongoda Mandala in the middle of the sixth century was of tribal origin as knwon from their inscriptions. The Cuttack Museum charter[26] mentions that once Kalinga was in a state of anarchy and an aboriginal chief named Pulindasena prayed to lord Sayambhu to save the country by providing an able ruler. Lord Sayambhu fulfilled his desire and created a luminous person out of the rocks, who was named Shailodbhava (*Saila-ud-bhava*) the progenitor of the dynasty. Similarly, the Bhaumakaras, who ruled Tosali region in the eighth century originally belonged to an aboriginal tribe called Bhuiyans (a variant of Bhauma) living in the hill tracts of Orissa.[27] Towards the later half of the fifth century AD, the Nalas who ruled over Koraput and Bastar region of Orissa and Madhya Pradesh respectively have claimed their descent from the Nisada King Nala of the epic fame.[28] King Sharabha, the progenitor of the Sharabhapuriya dynasty who came into prominence in South Kosala after the fall of the Nalas was of tribal origin.[29] The Bhanjas, who ruled over a number of Mandala states in several branches have mentioned that their family emerged from an egg of a pea-hen (*andajavamsha pravahava*).[30] This suggests that the Bhanjas originally belonged to some aboriginal tribe most probably the Bhuiyan, an ancient tribe of Orissa. The Shulkis who were ruling over

Kodalaka Mandala were originally tribals inhabiting the south-eastern parts of India.[31] The Varahas or the Mayuras of Bonai Mandala were originally from Chitrakuta,[32] the legendary home of the tribals and their family names *varaha* (boar) and *mayura* (peacock) suggest of the tribal origin. Similarly, many such instances can be cited to show the tribal origin of the ruling dynasties of early medieval Orissa.

Orissa being a part of the Eastern Ghats is primarily a mountainous, hilly region except the coastal plains. Most of the minor dynasties of early medieval Orissa ruled over Mandala states which were located in the forest tracts and hilly areas.[33] Also, the bigger dynasties such as the Shailodbhavas, the Bhaumakaras, the Somavamshis and the Later Eastern Gangas' area of control included the tribal dominated hilly, forest tracts. In those areas, the ruling chiefs donated a large number of land and villages to the Brahmanas. The Brahmana donees were mainly invited from outside to settle in the tribal hinterland of Orissa through land grants.[34] Such large-scale land grants accompanied by the transfer of various rights with immunities from taxation not only helped for royal legitimation in the newly emerged status in the tribal hinterland of Orissa but also for opening new areas for cultivation by introducing plough agriculture and peasantizing the tribals.[35]

Further, the newly emerged kingdoms in the upland tribals areas had no earlier experience of organized political system.[36] The most significant political development was that these small states became the 'nuclear areas' for diffusion of high traditions of Brahmanical culture. As Hermann Kulke has rightly stated, 'They were centres of integration of tribal elements rather than their "sustained displacement".'[37] The ruling chiefs of the tribal areas depended much on the loyalty of the local tribes to consolidate their political power. Sometimes, the tribals were adjusted in the caste hierarchy and given a particular rank in the political organization.

IV

The establishment of Brahmanic settlements through land grants, the spread of Vedic learning and construction of temples in the tribal areas not only led to the process of Brahmanization but also to the extension of agriculture which resulted in the economic development of the area.[38] Also, the ruling monarchs through the land grants brought the tribal chiefs into the Kshatriya fold.[39] Due to frequent wars and invasions the rulers of various dynasties recruited a large number of tribals into the army. This led to the upward mobility of the tribals with the formation of new sub-castes which were later on known as Khandayats and Paikas. They sometimes even claimed the status equivalent to the Kshatriyas and assimilated into the Brahmanic caste hierarchy in the process of Kshatriyaization.[40] They were patronized by the ruling dynasties and were assigned important military and civil offices. They assumed new titles such as *senapati*, *nayaka*, etc., and claimed higher status on the basis of their newly acquire landed estates and governmental position.[41]

Further, this process of Brahmanization facilitated the acculturation of the tribals. Also, some tribal elements were incorporated into the Hindu society. This interaction of tribal and Hindu culture led to the cross-cultural assimilation in Orissa. Many tribal deities, in early medieval times, were given Brahmanical sanctification and identified with the Shakti deities like Durga, Kali, Mangala, Chandi, Maheshvari, Pataneshvari, Khambeshvari, etc. These tribal deities may be variants of pre-historic fertility cults. They were related to the concept of female goddesses who protect men, ascertain the fertility of the female folk and soil, and accept blood sacrifice for her protection. They were patronized and adopted as *ishtadevata* (family deity) of the ruling chiefs.[42] Even now in Orissa many such tribal goddesses are being worshipped in different parts as presiding deities of a particular village, town or a political unit.

Another interesting example of the assimilation of tribal culture and Vishnuite Hinduism is found in the cult of Jagannatha which is the most important religious belief of the Oriya people. According to the tradition lord Jagannatha was originally worshipped by the aboriginal Shavara chief Visvabasu in the forest and which later on appeared at Puri.[43] The Jagannatha iconography still seems to be of a tribal look and the wooden images are of crude tribal origin. This fact is attested by the priestly class of Daitapatis who claim to be the descendants of the original tribal priests. This tribal religious assimilation is also reflected in the life and culture of the Oriya people.

V

This study, therefore, shows that from early times Orissa was inhabited by the tribals such as Shavaras, Pulindas, Shailajas, Gonds, Khonds, Bhuiyans, Bhils, etc., as known from the literary and inscriptional sources. There were a number of tribal principalities which sprang up in Orissa such as Aatavikarajya, Mahakantara, Mahendragiri, Kosala, Vidhyaatavi and Jhadakhandadesa and altogether they formed eighteen tribal principalities. A number of ruling dynasties of Orissa such as the Shailodbhavas, the Nalas, the Bhaumakaras, the Sarabhapuriyas, the Sulkis, the Bhanjas, the Varahas or the Mayuras and the Tungas have originated from the tribals and later on were upgraded to the Kshatriya status through Brahmanical association. Large number of land grants to the Brahmanas in the upland hilly forest areas of Orissa not only led to the process of Brahmanization but also contributed to the extension of agriculture in those areas. The land grants accompanied by the transfer of various rights with immunities from taxation helped for royal legitimation in the newly emerged states in the tribal hinterland of Orissa. Further, this process intensified a process of acculturation of tribals. Some tribal elements were incorporated with the 'Hindu' society and there was an upward mobility of the tribals which were assimilated in the Brahmanical caste hierarchy. Such interaction of tribal and Brahmanical culture led to the cross-cultural assimilation in Orissa as manifested in the worship of Shakti in form of mother goddesses and the cult of Jagannatha. Thus, the tribals have played a major role in shaping the history and culture of Orissa.

NOTES

* For recent editions of relevant epigraphs, see Ajay Mitra Shastri, *Inscriptions of the Sharabhapuriyas*, Panduvamshins and Somavamshins, Pts I and II (Delhi, 1995).

1. *Apadana*, ed. M.E. Lilley (London, 1925-7), Pt II, p. 350.
2. *Vinaya Pitaka I*, tr. T.W. Rhys Davids and H. Oldenburg (Oxford, 1881-5), p. 3; Mahavagga of the Vinaya text, 1st Khandaka, *Sacred Books of the East*, Vol. XIII, pp. 81-4.
3. *Vayu Purana*, Anandasrama Sanskrit Series (Poona, 1905), XLV, pp. 132-3; *Matsya Purana*, Anandasrama Sanskrit Series (1907), CXIII, pp. 52-3.
4. Op. cit.
5. *Majjhima Nikaya*, ed. V. Trenckner and R. Chalmers (London, 1888-1902), III, p. 78; *Anguttara Nikaya*, ed. R. Morris and Edmund Hardy (London, 1885-1900), II, p. 31.
6. Medhatithi, the Commentator of Manu, while referring to the Odras has remarked that the country derived its name from the tribe, cited by N.K. Sahu, *History of Orissa*, Vol. I (Utkal University, Bhubaneswar), 1964, p. 75.

7. *Natural History,* Eng. tr. (Loeb Classical Library edition, London and Cambridge, 1956-63), Vol. II, pp. 42-4.
8. *Ancient Geography of India* (Indian edition, Calcutta, 1924), p. 586.
9. Ibid.
10. *Aitareya Brahmana,* by A.B. Keith (Oxford, 1909), VII, 8.
11. Cited by S.K. Aiyangar, *Beginning of the South Indian History* (Madras, 1918), p. 160.
12. *Charjya Gitika,* ed. K. Mohapatra (in Oriya) (Cuttack, 1976), pp. 80, 83, 84, 99.
13. *Mahabharata* of Sarala Das, ed. A.B. Mohanty (Orissa Sahitya Academy, Bhubaneswar, 1968-70), *Sabha Parva,* p. 604.
14. N.N. Mishra, 'Tribal Elements in Jagmohan Ramayan', in K. Mahapatra (ed.), *Tribal Language and Culture* (Santiniketan, 1984), pp. 50-4.
15. See Rock Edict XIII, D.C. Sircar, *Select Inscriptions Bearing on Indian History and Civilization,* Vol. I (Calcutta, 1942), p. 35.
16. *Corpur Inscriptionum Indicarum,* Vol. III, pp. 6f.
17. *Journal of the Bihar and Orissa Research Society* (henceforth *JBORS*), Vol. XVII, Pt. I, p. 16.
18. Mahendragiri has been mentioned in several inscriptions of the Matharas, Shailodbhavas and the Gangas. See Dhavalapeta grant of Umavaram, *Journal of the Andhra Historical Research Society,* Vol. X, pp. 143-4; *Epigraphi Indica* (henceforth *EI*), Vol. XXVI, pp. 132-5. Mahendragiri is mentioned as one of the *Kulagiris* among seven *Kulagiris.* For details, see A.K. Rath, 'A Note on Mahendragiri', *Proceedings of the Orissa History Congress,* Jyoti Vihar Session, 1978, pp. 1-6.
19. See Kanasa Copperplate Grant of Sri Lokavigraha, *Journal of the Kalinga Historical Research Society,* Vol. II, No. 3, pp. 261-6.
20. S. Tripathy, 'Asanpat Nataraj Image Inscription', *Journal of Orissan History,* Vol. I, No. 2, pp. 5-10.
21. See Allahabad Pillar Inscription, ll. 19-20, op. cit.
22. Ibid.
23. The Polasara Plates of Arkesvaradeva, *EI,* XXVIII, pp. 63-7; *Inscriptions of Orissa,* Vol. III, Pt. II, pp. 354-7, l. 18.
24. *Journal of the Kalinga Historial Research Society,* Vol. II, No. 3, pp. 261-6.
25. The Bonai Copperplate of Vinita Tunga, *JBORS,* Vol. VI, p. 236.
26. *EI,* Vol. XXIV, p. 148. Also for details, see S.C. Behera, *Rise and Fall of the Sailodbhavas* (Calcutta, 1982), pp. 39-50.
27. B. Mishra, *Orissa under the Bhauma King* (Calcutta, 1934), p. 81.
28. N.K. Sahu, op. cit., p. 506.
29. Sarabha means of fabulous animal having eight legs stranger than a lion suggests tribal origin. See S.P. Tiwari, *History of Orissa* (Sambalpur, 1986), p. 68; Piparduha Copperplate Inscription of King Narendra, *Indian Historical Quarterly* (henceforth *IHQ*), Vol. XIX, pp. 139-46.
30. The Ukhundra Plate of Ranabhanja, *IHQ,* Vol. III, pp. 427-9. Two Adipur Copperplates of Narendrabhanjadeva, *EI,* Vol. XXV, pp. 147-61.
31. The *Markendeya Purana* describes the Shulkis as a tribe of the south-eastern India, see N.K. Sahu, P.K. Mishra, J.K. Sahu, *History of Orissa* (Cuttack, 1979), p. 145.
32. H.M. Mahtab, *History of Orissa,* Vol. I (Cuttack, 1959), p. 163.

33. For Mandala states, see N.K. Sahu, *Utkal University History of Orissa*, op. cit., pp. 110-29.
34. A.P. Sah, *Lie in Medieval Orissa, cir. 600-1200 A.D.* (Varanasi, 1976), pp. 119-20.
35. S.K. Panda, 'Landgrants and Extension of Agriculture in Early Medieval Orissa (A.D. 600-1200): A Study based on Epigraphic Evidence', paper presented to the national seminar on 'Aspects on Orissan Sculpture and Epigraphy', Utkal University, Bhubaneswar, 2-4 January 1997 (in press).
36. B.P. Sahu, 'Ancient Orissa: The Dynamics of Internal Transformation of the Tribal Society', *Proceedings of the Indian History Congress,* 45th Session, Annamalai, 1984, pp. 148-60.
37. H. Kulke, 'Early State Formation and Royal Legitimation in Late Ancient Orissa', in M.N. Das (ed.), *Sidelights on History and Culture of Orissa* (Cuttack, 1977), pp. 104-14.
38. S.K. Panda, 'Landgrants and Extension of Agriculture . . .', op. cit.
39. Surajit Sinha, 'State Formation and Rajpur Myth in Tribal Central India', *Man in India,* Vol. XLII, 1962, pp. 35-80.
40. H. Kulke, 'Kshatriyaization and Social Change: A Study in Orissa Setting', in S. Devidas Pillai (ed.), *Aspects of Changing India* (Bombay, 1978), pp. 398-409.
41. S.K. Panda, 'Social Mobility in Medieval Orissa', *Indica,* Vol, 21, No. 2, pp. 81-5.
42. H. Kulke, 'Tribal Deities on Princely Courts: The Feudatory Rajas of Central Orissa and their Tutelary Deities (*ishtadevata*)', in D. Sontheimer and M.L.K. Murthy (eds.), *Durga and Buffalo* (Wiesbaden, in press).
43. For tribal origin of the cult of Jagannatha, see B.M. Padhi, *Daru Deata* (in Oriya), (Cuttack, 1964); K.C. Mishra, *The Cult of Jagannatha* (Calcutta, 1971); R. Gieb, *Die Indradyumna Legende, Ein Beitrag zur Geschicte des Jagannatha Kultes* (in German), (Wiesbaden, 1975); A. Eschmann, 'Hinduization of Tribal Deities in Orissa: The Shakta and Shaiva Typology', in A. Eschmann, H. Kulke, G.C. Tripathy (eds.), *The Cult of Jagannatha and the Regional Traditions of Orissa* (New Delhi, 1978), pp. 79-97.

Adivasi and *Aranyaka*: Reconsidering Some Characterizations of Their Polity and Economy in Pre-colonial and Colonial India

B.B. Chaudhuri

Aranyaka (forest-dwellers and forest users) and *Adivasi* (tribes) were not necessarily exclusive categories. They were not identical either. *Aranyakas* did occasionally include *adivasis*, but not all *aranyakas* were *adivasis*.

While distinguishing between them I am not taking an essentialist position. Their formation was historically determined, and their attributes were not, therefore, fixed forever everywhere. The word '*adivasi*' need not be taken in its literal sense, i.e. original settlers or aborigines. Migrations for various reasons commonly occurred and new migrants as a result pushed out old settlers. We may take as an instance the encounters between the Mundas and Oraons, the two most numerous *adivasi* communities of south Bihar. The Mundas constantly retreated in the face of Oraon intrusions into their village settlements. 'Tribe', however, had a few markers, especially where the subsistence was mainly based on settled cultivation. A notable one was clan or kinship-based agricultural settlements here, in the sense that a clan or a kinship group took the decisive role in these settlements and in the foundation of villages. (The clan or kinship could grow out of the original village founder's family.) The descendants of the village founder's family had normally the prior access to village resources, and the social organization that developed around the settlements sought to ensure it. The chief functionaries of the village organization, its headman and priest, also belonged to the family. The later settlers in the village and also those not connected with the family on the father's side had inferior rights in the village land and other resources. The original settlers carefully maintained their separateness from the later settlers, especially those speaking a different language.

Aranyaka formed a composite group. It included four main communities: settled peasants living close to forests and using forest resources, occasionally land; tribal groups practising settled cultivation and also using forest resources; pastoral and nomadic communities and shifting cultivators. The communities did change over time their modes of subsistence, for instance, change from shifting cultivation and pastoral occupation to settled cultivation. Hence the absence of any fixed boundaries between them. In view of the increasing attention given to forest communities, as distinguishable from settled peasants, in the recent studies on the impact of colonial rule on rural India, we need to be cautious against unduly stressing the distinction between the two communities. At least from the point of view of peasant enterprise in increasing cultivation, settled cultivation and forests need not be taken as distinct zones of economic activity. In the perception of peasants and also of adivasis practising settled cultivation, forests, especially those near their villages, were as much a part of their means of subsistence as settled agriculture was. This was perhaps the reason why the 'measured area' (*arazi*) of Mughal surveyors included gross cultivation, cultivable waste and part of the uncultivated land.[1] A recent work on the agrarian economy of Tamilnadu in the early nineteenth century makes a similar point. 'Waste has been treated as a part and parcel of local culture, often connected with the changing agro-ecosystem of the locality. . . .' The Tamil word for waste *tarisu* was 'used to mean cultivable waste, which was distinguishable from what is called *purambaku* or land incapable of cultivation'.[2] Again, as we have just said, change from one mode of subsistence to another often occurred among forest communities.

We first examine two characterizations of the pre-colonial economy and polity of the *adivasi* and *aranyaka* societies. The first relates to the nature of forest control and forest use by pre-colonial rural communities and governments, and the second to the nature of pre-colonial tribal village's links with the wider polity and economy and their implications for the tribal village organization. We then briefly indicate our approach to the study of the nature of colonial impact on these societies.

II

We first examine an argument of the pioneering study of Gadgil and Guha[3] relating to the nature of control and use of forest resources in pre-colonial times. Two broad types of the use over time have been identified: 'prudent' and 'profligate' (wasteful and 'rapacious'). While the use in pre-colonial times, generally 'prudent' prevented damage to the ecosystem and ensured ecological balance or equilibrium, it tended to be increasingly profligate during colonial rule. 'British colonial rule marks a crucial watershed in the ecological history of India.'[4]

There are two explanations of why the pre-colonial use was prudent: 'cultural tradition of prudence' at the level of the community, called 'conservation from below', and the particular use and control of forest resources at the level of the state (the Mughal state in this case). The first argues that the village community followed a collectively recognized set of restraints in using forest resources, thus precluding their 'rapacious' use and ensuring 'the long-term viability of production'. As we presently see, the determinant of the restraints was partly communal religious beliefs, rituals and practices. The institutional mechanism that also made possible this 'conservation from below' was the caste system and the manner of its organization. The endogamous groups composing the village caste society 'tended to pursue a relatively specialized and hereditary mode of subsistence'. This society, having 'mutually supporting relationships' among the groups 'developed a variety of institutions to regulate the use of living resources'. The village society devised here 'an elaborate system of the diversified use of the living resources that greatly reduced inter-caste competition'. As a result, 'any particular resource of a given region used to be utilized by a small homogeneous caste group, which expected the same resource to sustain its future generations also'. The second explanation is inadequately presented. 'Even the Mughals, whose religion was Islam, were unable perhaps unwilling to alter the existing pattern of resource use and the social structures in which they were embedded.' The general argument is that the needs of the

Mughal state differed from those of the colonial state. Why the former did not lead to any change in the 'existing pattern of resource use' has not been clearly stated.

The evidence of the 'cultural tradition of prudence' usually cited is some communal practices connected with religious beliefs: the collective choices of a village site for worship, generally the shade of a cluster of trees. The practice is more a characteristic of tribal villages than of settled Hindu peasant villages, where the place of worship was mostly a temple. Since the village site chosen for worship was a 'sacred' place, felling of any trees in the cluster, even cutting of branches of tees, was forbidden. Tribals believed the tree cluster was one of the abodes of their gods, especially 'village gods', distinguishable from the gods of the tribe as a collectivity. Tree felling was, therefore, an act of blasphemy, since it would disturb the gods. Offenders guilty of the 'profane' activity had severe penalties to face. Villager seldom violated the social injunction.

This 'cultural tradition' did not, however, have the kind of ecological implication that Gadgil and Guha supposed. It just meant *exclusion* of a specified village space from cultivation forever, and not its 'prudent' use from time to time. The exclusion had nothing to do with tribal/peasant village's attitude to nature and with any ecological concerns. Even this 'tradition', sentiments of reverence for the sacred tree cluster, weakened, at least in the Munda and the Ho villages of south Bihar, in the later years of colonial rule. Zamindars having proprietary control over the villages seldom respected the tradition. Even the villagers themselves became irreverent towards it. Trees were secretly felled, though not for the purpose of cultivation. The 'cultural tradition of prudence' did not, however, work in any part of the village space outside the sacred site. For instance, there was no ban on new cultivation out of forests elsewhere, or on the use of forests for other purposes. Iron smelting, where the village had no other easy source of supply of iron, a vital need for its economy, required a big supply of firewood as a source of fuel. This inevitably damaged forest resources.

The scope of 'communal control' that the notion of 'cultural tradition' assumes did not necessarily exist everywhere. Writing of the Himachal region, Chetan Singh has even doubted the existence of 'village community' in its clear-cut and definite sense.[5] Singh does admit that 'in pressing matters requiring immediate resolution at a village level a broad consensus was quite often easily reached by peasants'. Habib finds 'community' a valid notion in the sense of a 'corporate body for collective action', 'mostly when it confronted the state machinery'.[6] The 'community' sentiments, Chetan Singh points out, had also cultural roots. 'The role of the local deota (god) was quite significant in these matters.' The sense of the 'community' was, then, perhaps the strongest among villages sharing a common religious faith. The 'village community' (implying some form of communal control) was also a valid concept mainly in the sense that the community could control use of the existing resources, especially in order to prevent free access of outsiders to them. For instance, villagers connected with the village founder's family enjoyed some special privileges in regard to landholding, and they could not transfer their holdings to outsiders without the community's prior consent. Such transfers did occasionally occur, but the communal control could normally limit intrusion of others. This was particularly true of tribal villages. A village, with more or less recognized boundaries, did include the resources of forests close by it. Whether the village community could also control the resources of forests beyond its recognized limits depended on who actually had authority on them.

Two things are notable here. Apart from the fact that reverence for the sacred village site invariably derived from religious beliefs, peasant or tribal villages, while keen on increasing cultivation, had no recognizable means of knowing if going beyond a limit could in the long run damage the locality's ecosystem. Communal control in regard to the use of 'living resources' mainly related to the question of access to the resources, and not to its 'conservation'. Secondly, the precise scope of 'communal control' needs to be carefully determined. Much depended on the presence of groups, within and outside the village, whose interests conflicted with the interests of the village.

Far less convincing is the second argument of Gadgil and Guha. Caste system did mean existence of distinct occupation groups, and as long as they remained distinct the use of the village resources was non-competitive. However, the idea of non-competitive occupation groups need not be overstressed. Modes of subsistence of different social groups seldom remained fixed for long. Movement from one occupation to another frequently occurred. Even settled did not at all mean that cultivators did not ever move out of their original settlements. Peasant mobility, Habib shows, was a common occurrence in Mughal India.[7] More relevant for the purpose of our study is whether the use of forest resources by groups directly or partially involved in it was non-competitive in nature. The nature of inter-occupation mobility here differed. Some cases more commonly occurred than others. Less common was the movement from settled cultivation to cultivation in forests, generally shifting cultivation. This generally resulted from the migration and agricultural settlements by more powerful groups who succeeded in pushing out the earlier settlers. This happened, for instance in parts of what came to be called Central Provinces. Peasant settlers here, unable to cope with the might of politically resourceful groups, retreated to the sparsely inhabited hills taking to shifting cultivation as their living.[8] A more common thing was the gradual replacement by migrants skilled in wet farming of shifting cultivators tied to an inferior farming technique. This was how the Santals pushed out the aboriginal *Paharias* in the Rajmahal Hills of Bihar. Also common were changes of pastoral/nomadic groups into settled cultivators. All this is evidence of a continuing contest between different groups over the use of forest resources in different forms.

Too much of emphasis on 'community' and on the distinctiveness of caste-based occupations tends to ignore the class relations in the village, especially non-tribal villages that determined the use of forest resources. Interests of zamindars, for instance, did not necessarily harmonize with the interests of the village communal collectivity.

We, therefore, need to look beyond the village and consider the role of supra-village dominant groups including the state. Two things need to be examined here: first, Gadgil/Guha's conclusion that the Mughal state's practices did not alter the 'existing pattern of resources' and secondly, an explanation of the continued existence of certain 'rights' of forest dwellers in pre-colonial times.

'Existing pattern of resources' is a vague phrase. The 'pattern' needs to be identified. The implications of the state's practices for the ecosystem and for the forest dwellers' modes of subsistence have not been quantified. What we are able to do is to describe the different uses of forest resources by the state and to suggest the possible implications. Three major needs prompting the state to use forests or to control them were increased income from new cultivation out of forests, needs of a political and military nature and royal hunting requiring control over a big area abounding in games.

Only the second need had a direct bearing on the ecosystem. It had a dual role. It sometimes led the state to preserve the forest where it judged that its density obstructed enemy movement. More often, the need involved destruction of forests. A secure control over forests was an effective means for the state to flush out law evaders and political rebels, including zamindars, who often hid there. Its overriding necessity obliged the state to adopt any measure, including burning of forests or indiscriminate felling of forests, unmindful of its ecological effects. Military necessity also partly necessitated construction of roads passing through forests and reinforced the need for state control over them. The manner in which the new cultivation was done, which the state indirectly encouraged mostly through offers of liberal terms of revenue for a period of time, strengthened the regional economy and also the base of the peasant society. We have no idea about whether the reclamation process involved sacrifice of the interests of groups such as shifting cultivators. (The colonial government did not mind the sacrifice presuming that shifting cultivation was a wasteful mode of cultivation.) Hunting, a common pastime of emperors and nobles, did require that outsiders did not poach on the hunting zone. Royal hunting too did not normally damage the forests. It only restricted, sometimes temporarily, outsiders' access to the royal preserve. Hunting,[9] Rangarajan concludes, 'was important for several Indian rulers, but the scale of intervention in regulating access to jungle was limited. There is no evidence of a consistent pattern of exclusion, except within certain reserves set aside for the use of rulers or nobles.'

Some scholars argue commercial logging—felling of trees with an eye to trade in timber—was not at all a new thing in colonial times. 'Indian regimes had sought to tax and monopolize valuable timbers.'[10] The teak forests of Travancore were thus a source of considerable income for the King Martanda Varma. Study of some 'pre-colonial deforestation episodes' led Richard Grove to question the characterization of 'the pre-British period as an ecological and pre-capitalist golden age of common property rights and sustainable resource use'.[11]

However, scholars generally agree that the scale of the logging was too small to cause any widespread ecological damage. The Mysore Kings' trade in timber was wholly restricted to sandalwood. Considering the uses of the timber at the time, the number of trees felled was far smaller compared to the colonial times when the pressing and growing need for timber for railways, shipbuilding and urban housing necessitated logging on a vast scale.

We also need to explain why the relatively easy access of forest-users in many a place to forest resources—such as land, grazing space and forest products, continued. This was not due to any recognized principle of statecraft. The access was far from universally permitted. Where the access prevailed, the reasons were that the state did not find it inconsistent with its needs, or that it did not detect it, or even where it was detected, it did not have the right means of preventing it. The fact that the activities of the government, except in the cases noted above, had generally a limited role in damaging forest resources did not, however, necessarily mean the village community's free access to them.

In conclusion, we do not see any conscious designs of the village community or the state towards preserving the ecosystem. If the ecosystem still continued undamaged, the cause was the particular nature of the uses of forests by the village community and the state. Forests were depleted for several reasons, but the scale of the depletion was not big enough to disrupt the old ecological order.

An argument of Gadgil/Guha and others highlighting the ideological roots of the colonial state's policies and practices also needs to be examined. The argument emphasizes the influence on the colonial bureaucracy of the contemporary economic teachings in England and in Europe in general. The teachings reflected the influence of the qualitative transformation of the economy of Europe as a result of the Industrial Revolution. The Revolution involved a gradual transformation of natural resources into commodity because of their use in the industrial process. In the sphere of the agricultural sector of the nation's economy, it also meant valorization of settled peasant cultivation as a rational use of natural resources. Any system that involved wastage of the resources was deprecated. This was one reason why the colonial state regarded, especially initially, the continued existence of forests as a mark of economic backwardness and the expansion of cultivation as a mark of civilization and progress. Reduction or elimination of wild spaces, i.e. forests was considered par of a 'civilizing mission' by it.[12] The teachings, it is also argued, had much to do with the state's insistence on replacement of shifting cultivation and pastoral-nomadic economy with settled peasant agriculture. A general conviction that a big 'floating population' engaged in these two economic systems was a potential threat to the 'law and order' and that a 'sedentary population' formed a sound base for political stability reinforced the state's aversion to these two modes of subsistence.

The teachings did influence colonial bureaucrats in the sense that they were part of the latter's intellectual and cultural background. The influence need not be exaggerated. (Ajay Skaria read too much into the occasional official pronouncements opposing 'civilization' to 'wildness'.) It would be wrong to infer that the teachings created an altogether new trend in the formulation of the agrarian policy. (An example of a novel trend was the influence, though short-lived, of the utilitarian theory of rent on the 'revenue settlements' in the Bombay Presidency and the North-Western Provinces.) The teachings do not explain the colonial state's insistence on increase in cultivation. The reason was a simple one. Land revenue formed the major source of the state's income and its size depended on the state of agriculture. This was not unique to the colonial state. This was also the reason why the Mughal state was keen on expansion of cultivation and the security of peasant producers as a sound base for this. 'Its political philosophy', concludes Chetan Singh, 'similarly valorized a fixed peasant population and distrusted pastoral-nomads, a group ever on the move.' 'For the Mughals a civilized society was one primarily engaged in agriculture, alongside more sophisticated commercial activity. It was nurtured by . . . a state that regarded protection of the peasant as an essential obligation and extension of agriculture as its cardinal objective.'[13]

The colonial state was merely continuing an old norm of statecraft. The economic teachings did not lead the state to plan organizing the agricultural system in a new way, for instance, through investment in new cultivation or in increase in the productivity of the old. An exception was investment in irrigation. This was, however, intended as a security of the existing cultivation, threatened from time to time by droughts ruining crops.

The teachings did influence the government attitude to the relatively backward agricultural systems that impeded growth of cultivation and caused wastage of potential agricultural wealth. This was the reason why the state wanted shifting cultivation replaced with settled cultivation. It presumed shifting cultivation was wasteful because its slash-and-burn method destroyed forest resources; it was inefficient because it precluded intensive cultivation and increased productivity. The influence need not be unduly stressed. In brief, though the government banned the system, the formal ban did not lead anywhere to a substantial increase in settled cultivation. Secondly, the ban was more effective in restricting the area under shifting cultivation and in using the forests thus retrieved for a very different purpose: income from the sale of forest products and from renting the land now controlled by the Forest Department.

An explanation of the Government's dislike of shifting cultivation in terms of its political philosophy that settled agriculture conduced to a stable law and order is only partially valid. Government's initial practices were at variance with it. To the extent that settled cultivation *normally* ensured a secure crop, it minimized the chances of 'disorder' generally following a widespread crop failure. Raids on the plains were a common 'crime' during acute scarcities caused by crop failures in the hills. This threat to 'law and order' particularly worried government when landlords and others in the plains used the raiders to promote their own interests, for instance, in their conflicts with their local rivals. Settled cultivation, the government believed, would minimize the threat.

Government's *initial* practices show that it did not invariably regard settled cultivation as a means of coping with the threat to the 'law and order'. It did not mind preserving even an archaic agricultural system where it felt its abrupt disruption or breakdown would create new threats to local peace and order. This is evident from its attitude to the *Paharia* practice of shifting cultivation (*curao)* in the Rajmahal Hills of Bihar. This traditional system came under threat with the Santal immigrants, skilled in wet farming, increasingly poaching over the *Paharia* lands. Worried over the resultant clashes between them, government sought as a solution to contain the Santal advance, initially by marking off the *Paharia* domain and forbidding the Santals to cross over. At the same time it encouraged the *Paharias* to learn the art of wet farming, so that they could compete with the Santals. All was in vain. There was no stopping the Santal intrusion. Nor did the *Paharias* change their cultivation method. They retreated further up the hills. Government attitude changed before long. Ever increasing new cultivation through Santal enterprise convinced government that it made no sense artificially propping the *curao.* Repeated *Paharia* appeals to the government to come to their rescue went unheeded.

Ideological influences on the colonial state's forest policy need not, therefore, be overstressed. The decisive influence was its needs, which significantly differed from those of the earlier governments. Ideology only reinforced its conviction of the worth of the needs. It could also more confidently enforce its claim with the increasing refinement of its apparatus of rural control since about the late nineteenth century.

The state's forest policy evolved over time with its changing needs. Possessing far greater powers it could also enforce its claim. Hence, the deeper impact of the policy on the region's ecology. The government did not work throughout with a fixed set of assumptions. A result was important shifts in its policy from time to time. We notice three major shifts. First, badly needing increased revenue income government initially regarded control over forests, an ideal site of new cultivation, as an important source of it. The second shift occurred when with the increasing need for timber, first for shipbuilding, in India and in the metropolis,[14] and to a larger extent later, for sleepers in the fast growing railways, forests acquired a new value and their importance as a source of new cultivation declined. 'Preservation' of forests marked the third decisive shift in the state's policy, and aspects of it deeply affected the regional economies. The case for 'preservation' of forests only partly rested on immediate pecuniary calculations. One major concern, *at least initially*, was ecological. Government then believed deforestation, apart from causing loss of precious resources, would reduce rainfall, harm the topsoil and tell on productivity of land. The policy also differed in different provinces to suit local conditions. Sivaramakrishnan has shown how government's perception of what it should do and what it eventually did was influenced by its interaction with the local polity and economy.[15] We ignore here such shifts and concentrate on the major trends in the policy. We emphasise how the colonial state's needs, and its measures to meet them substantially differed from the pre-colonial state's practices, and in the process deeply impinged on the forest dwellers' economies.

III

The characterizations of the *adivasi* society that we would now examine relate mainly to a crucial aspect of it—its involvement and, at times, integration with the wider polity and economy in pre-colonial and colonial times. Scholars have generally agreed that the *adivasi* world had not ever been a 'primitive isolate'. The reason of our emphasis on this aspect of the *adivasi* society is our hypothesis that it was its increasing involvement with the larger unit that often determined the long-term process of structural change in it. (We use the word 'structure' in the sense of a relatively enduring relationship of its component parts.) We discuss the question with particular reference to the tribal world of the 'permanently settled' part of what came to be called the Bengal Presidency. We have brought in the history of other regions mainly by way of comparison.

First we see how *adivasi* villages became linked in different ways with the non-tribal polity in the neighbourhood. Four of them are particularly significant. We then consider characterizations of some of them.

A common one, which did not necessarily lead to any stable political control over them, was their unavoidable dependence on peasant neighbours for basic subsistence needs. Sometimes they learnt aspects of the art of cultivation. What they borrowed were not generally tools of cultivation, but the knowledge that contributed to security of agriculture, for instance, knowledge of the right kind and time of cropping. *Adivasis* could not also help depending on neighbours for other needs that could not be wholly internally provided such as cloth, iron and credit. Iron was a vital necessity for the making of ploughs and other tools. *Adivasi* villages having no iron ores drew on external supplies. The biggest source of credit was external too where the tribal village surplus was too small for the creation of the savings that could be used as credit. (This was the reason why non-tribal artisan communities, such as weavers and blacksmiths, had a prominent position in the tribal village organization.) The army and police of powerful zamindars in the adjacent villages were also a source of their employment. (They served as *paiks* and *ghatwals* in several parts of south-western Bengal.) Structural insecurity of *adivasi* agriculture, especially due to inadequate water supply and the consequent dependence on a single crop, often led them, in times of crop failures, to raid the plains. In this they generally counted on the protection of the powerful zamindars spoiling for a fight with their rivals. This was how the *Paharias*—called 'hillmen' in contemporary official records—of the district of Birbhum came to the involved in the inter-zamindar clashes there. The history of Santal agricultural settlements also shows how the Santals, constantly on the move in search of new land, settled in zamindari domains, and also where powerful non-tribal chiefs had established control over pre-existing tribal settlements.

Second, some recent studies[16] show how the tribal village chief's employment in the administration of the neighbouring governments formed a vital link in the tribal village's 'interaction with larger civilizations'. An important finding of Sumit Guha's inquiry, 'focussed on the margins of agriculture in Central and Western Maharashtra', is how 'forest folk . . . partook prominently of one of the mainstream cultures of medieval South Asia—that of soldier rather than that of the cleric, merchant or peasant'. 'Certain' 'implicit' assumptions of his 'about the understandings and aspirations of these peoples and their chiefs' include 'their ability and desire to use their habitat, knowledge and skills to secure integration at the warrior level of the social hierarchy'. Many of the 'legends' of the Bhils 'suggest aspirations towards warrior status on their part'.

A comparable case was the role of some Gond tribal village chiefs (in the Central Provinces). Baker's study of the origins of the role is suggestive. Originally a village headman the chief gradually acquired powers which did not just derive from his role as the headman. One possible source was his new cultural status that his conversion to Hinduism gave him. This at least elevated his social and political position in the eyes of the Hindu kings in the neighbourhood. It is not known if Gond villages, at least some of them, owed also to the Hindu influence their superior agricultural technique, especially plough cultivation, which was widely in use in the Hindu villages. Whatever the reasons for the adoption of plough cultivation method, its replacement of the usual tribal practice of shifting cultivation made cultivation more efficient and made the villages more resourceful. That the chiefs continued to be powerful in the village is evident from their choice as *malguzars* by Richard Temple's administration in the Central Provinces (in the 1860s).

Nandini Sundar's study[17] of the *adivasi* world of Bastar in the Central Provinces shows a more broad-based form of the tribal village's involvement in the 'larger civilizations'. Among other things it brings out the complex cultural relationship of the *adivasis* with the wider Hindu culture, the chief protagonist of which was the royal family. As elsewhere, *adivasi* emulation of Hindu culture occurred here too. What is striking in Sundar's study is how the *adivasi* cultural mores also affected Hindu religious beliefs and rituals. For instance, Danteswari, the presiding deity of the royal family, was originally a tribal goddess. Furthermore, this helped the royal family legitimize its authority over the *adivasis*. As evidence of the royal family's success in this Sundar analyses the forms of the *adivasi* village's participation in the celebration of Dussehra, the chief festival of the royal family.

A distinctive form of the tribal village's *integration* in the wider political unit occurred through what is called the process of 'state formation', i.e. the rise and growth of a central political authority from *within* the *adivasi* world itself. (A view argues against classifying tribes as 'primordial communities', and regards them as 'secondary formations, i.e. reactions to the formations of states'.[18] The point we argue is different. Even if we assume that tribes were 'secondary formations', they could well exist without being integrated in the states recently formed.) While at the time of its origins the authority was not external to this world, it became increasingly autonomous of its tribal roots. A royal family, non-existent earlier (for instance, the Nagbansi raj in the Munda/Oraon region of Bihar) exercised the authority and increasingly distanced itself from the *adivasi* society. The process of Hinduization of the royal family gradually completed its alienation from the *adivasi* fraternity. The royal family's conversion to Hinduism barely affected the traditional religious faith and rituals of the Munda-Oraons. On the other hand, it was the tribal villages that mostly paid for the maintenance of the state machinery.

Of these four types of contacts of the *adivasi* village with the outside world we leave out the first. To what extent did the rest actually lead to the integration of the adivasi collectivity in the 'larger civilizations'?

Sumit Guha's emphasis is on the role of the tribal village chiefs in 'the contemporary regional politics', especially in the army. This role, Guha argues, fulfilled the 'aspirations' of these 'forest folk' and 'their chiefs', and contributed to their 'integration at the warrior level of the social hierarchy'.

What exactly were the implications of the chief 's role for the tribal village social organization. To what extent did this relationship, apart from the chief 's enhanced political and economic status, affect the tribal village organization? Guha has not looked at the question from this angle. We need to know who these chiefs were, their origins, the precise nature of their role in the tribal village organization and whether the chiefs came to be involved in the neighbouring political system as individuals and not as representative of the tribal village community. Where they came as individuals, the tribal community would not be affected unless their enhanced resources now led them to claim a superior position in the village organization. The chiefs were perhaps originally village headmen, enjoying thus a special position in the village. This presumably had much to do with the decision of the regional governments to employ them.

Nandini Sundar's study of the *adivasi* world of Bastar enriches our understanding of two levels of its interaction with the 'larger' Hindu civilization. *Adivasis* emulated traits of Hindu religious beliefs and the royal family assimilated into the Hindu pantheon tribal gods and goddesses. As an illustration of the family's 'legitimization' of its authority over the *adivasis* she refers to their eager participation in the celebration of Dussehra, the most important of the royal family's festivals.

Several things are notable in this regard. First, Dussehra celebrated in the region for long did not remain just a Hindu festival. It was a time of gaiety for a larger collectivity, including the *adivasis*. (This was also the reason why Birbhum Santals took part in the Durga Puja celebration in the locality. It is misleading to argue that the celebration was a unique event only during the Santal *hul* of 1855. Contrary to a point of view, it was not in the least connected with Santal attempt to appropriate aspects of the culture of the dominant Hindus as a means of subverting the latter's cultural domination.) Second, the forms of the *adivasi* celebration of Dussehra were inseparable from the old practice of unpaid labour service (*betbegar*) claimed by the royal family from the *adivasis*. This was just an evidence of the family's entrenched control over the *adivasi* village's resources, including labour. Third, the so-called 'legitimization' of the royal authority had a fragile base. Signs of intrinsically antagonistic *adivasi* relations with the ruling family increasingly appeared, especially since the later decades of the nineteenth century.[19] The undercurrent of this deep resentment towards the family amply shows that *adivasis* seldom accepted the political culture of rulers. The distinctiveness of the *adivasi* society and culture was partly reflected in the way *adivasis* and their Hindu neighbours viewed each other. Hindus regarded *adivasis* as being alien to their culture. Tribal culture was stigmatized as an inferior type. (The colonial state appropriated this concept.) *Adivasis* too boasted of their separateness from arrogant Hindus, contrasting their manliness, especially their robust physique, with the timidity of Hindu weaklings. In his study of the Dangs of Maharashtra, based partly on the Dang oral tradition, Ajay Skaria has shown how the Dangs took pride in their 'wildness', regarding it as an unique thing marking them off from the Hindu caste hierarchy.[20]

'State formation', i.e. the rise and growth of a centralized political authority from within the tribal world itself, was the most enduring, decisive form of the *adivasi* village's integration in the larger polity and economy. The point of our emphasis is that though the process of state formation significantly modified the old tribal village organization, it is misleading to conclude that it had disintegrated even before the colonial intervention. Colonial rule formed the decisive break. Our empirical evidence mostly relates to the tribal world of the permanently settled part of the Bengal Presidency, Bengal and Bihar. It was the Nagbansi raj that represented this centralized political authority in the Munda-Oraon region of south Bihar.

The impression that the old tribal order had already disintegrated before the colonial intrusion is widespread among scholars.[21] Rise of the Nagbansi raj did impinge on the *adivasi* world. The villages had to surrender part of their resources, land and labour, towards the maintenance of the state machinery. The apparatus of its control meant entry of a horde of non-*adivasi* aliens, including merchant-financiers dealing in luxuries and costly commodities. The raj family thought their consumption befitted the royal grandeur, especially after the royal family started emulating the lifestyle of the Mughal emperors and nobles. The ruling family barely controlled the activities of the aliens. However, though the state claimed part of the tribal village surplus, it did not interfere in the rural economy in the way the colonial state did, as we would presently see. The raj did not intervene either in the tribal village organization, for instance, in the activities of its chief functionaries, including the village headman (*munda*) and the village priest (*pahan*). The Nagbansi seldom bothered to impose Hindu religious norms on the *adivasis*.

It was with the colonial rule that the decisive break with the past occurred. We notice five major phases in it, though there are some overlaps. The first is from the beginning of penetration of colonial control into the *adivasi* world to the outbreak of a series of revolts here in 1830-2. Three major things happened during the period. The colonial state then consolidated its authority at the cost of established power groups—chiefs and others, which were distinguishable from the functionaries of the tribal village community. The second was the widening of the administrative frontier necessitated by the process of this consolidation. The economic frontier too widened then, primarily in the form of intervention of an essentially mercantilist state, wholly averse to productive investment in the economy, but keen on its regulation through administrative fiats, as a source its increased income.[22] The second phase ends with the Santal revolt (1855). The partisan intervention of the colonial state in favour of the dominant alien groups in the tribal village then crystallized. The third phase covers the years 1856-1900 (1900 was the year of the Munda movement led by Birsa Munda). The state policy then further helped alien domination consolidate itself. The fourth phase, which partly overlaps the second and third, marks the deepening of colonial encroachment on the forest resources that had a crucial role in the *adivasi* economy. In the last phase the government sought to cope with the difficult situation in the *adivasi* world since the suppression of Birsa Munda's movement. It was also worried over the new types of *adivasi* uprisings in the first three decades of the twentieth century, some influenced by the nationalist ideology. It was also then that we notice a shift in its policy regarding *adivasis*. Bureaucrats (and also others concerned over *adivasi* welfare) did not agree on the means of checking once for all alien intrusion into tribal villages, which was generally believed to be the major cause of their ruin. One view favoured closing of the *adivasi* world to aliens. This, it was also felt, would help the distinctive tribal culture and civilization survive. (Anthropologists then debated the question of the place of tribal society and culture in the mainstream Indian civilization. Nationalists assumed their organic relationship.) Others thought government action cannot do anything about the deep-rooted alien control.

All these developments led to the breakup of the old tribal village organization. At the same time *adivasi* subsistence was severely threatened. The two processes of change created a pervasive sense of frustration and helplessness in the *adivasi* world because of the steady erosion of its faith in the state as a source of protection and justice. This formed the background to the gradual radicalization of the ideology and organization of *adivasi* revolts.

In conclusion, the breakdown of the village communal organization had little to do with any internal changes in the tribal society and economy. It was largely due to exogenous factors, all connected with the intrusion of aliens into the tribal village, including the colonial state itself. Aliens did settle in *adivasi* villages earlier too, but the villagers carefully distinguished between what happened earlier and what happened in colonial times. The Bastar Dangs, as Ajay Skaria tells us, called the pre-colonial times Mughlai and the colonial period Mandini. 'Mughlai' was the time when the *adivasis* could freely move about in the forest. 'Mandini' was the historical epoch associated with British dominance, when this freedom nearly disappeared. I would here emphasize how the state's severe anti-insurgency measures most severely damaged the *adivasi* village organization. The state's disdainful attitude towards *adivasis*, evidently influenced by the Hindu concept that they belonged to an inferior civilization and culture, partly accounted for the mindless cruelty against the rebels. The markedly partisan role of the state immensely strengthened the adversaries of *adivasis*. The increasingly arrogant assertion of power of the Munda royal family in dealing with the Munda is an instance of this. This attitude of the state also generally encouraged the alien land-controllers gradually to eliminate the privileged members of the village founder's family. The link of the village headmen with the communal village organization became tenuous as a result. The state and the aliens ceased to regard them as functionaries of the village community, and used them as mere collectors of rent and other dues from the village. The weakening of village headmen's powers and the decisive role of the state in deciding village disputes, etc., disrupted the internal functions of the village society. This was partly evident in the increasingly bitter relations of village headmen and village priests. A new irritant in the *adivasi* village was the increasing arrival here of cultivators from the neighbourhood. Regarding them as more skilled agriculturists, alien proprietors and also holders of short-term leases of estates lured them to come over. Another inducement for them was the gradual opening up of the *adivasi* world through expansion of trading network and better transport. Earlier, the *adivasi* village did admit outsiders on its own terms because of their positive role in its economy.

NOTES

1. Shireen Moosvi, *The Economy of the Mughal Empire c.1595: A Statistical Study* (OUP, Delhi, 1987), pp. 44-7.
2. Arun Bandopadhyay, 'Environment and Identity in the Indian Historical Context: An Initial Vision', in Bharati Roy and David Taylor (eds.), *Politics and Identity in South Asia* (K.P. Bagchi & Co, Kolkata, 2001), pp. 144-7.
3. Madhav Gadgil and Ramachandra Guha, *This Fissured Land: An Ecological History of India* (OUP, 1992), Pts Two and Three.
4. Ibid., p. 5.
5. Chetan Singh, *Natural Premises: Ecology and Peasant Life in the Western Himalaya, 1800-1950* (OUP, 1998), pp. 208-9.
6. Irfan Habib, *The Agrarian System of Mughal India, 1556-1707* (OUP, Delhi, 1999), 2nd edn, p. 144.
7. *The Agrarian System* . . . , pp. 20-1, 133-5.
8. D.E.U. Baker, *Colonialism in an Indian Hinterland: The Central Provinces, 1820-1920* (OUP, Delhi, 1993), pp. 79-80.
9. Mahesh Rangarajan, *Fencing the Forest; Conservation and Ecological Change in India's Central Provinces, 1860-1914* (OUP, 1996).
10. C.A. Bayly, *Indian Society and the Making of the British Empire* (CUP, 1988), p. 140.
11. Richard Grove, *Green Imperialism: Colonial Expansion, Tropical Island Edens and the Origins of Environmentalism, 1600-1860* (OUP, Delhi, 1995), pp. 386-8. As illustrations he refers to the activities of the Rajah of Cochin, the Amirs of Sind, and the 'Maratha state builders'.
12. Also see Ajay Skaria, *Hybrid Histories: Forests, Frontiers and Wilderness in Western* India (OUP, Delhi 1999), 'Preface'.
13. Chetan Singh, 'Forests, Pastoralists and Agrarian Society in Mughal India', in D. Arnold and R. Guha (eds.), *Nature, Culture , Imperialism: Essays in the Environmental History of South Asia* (OUP, Delhi, 1995), pp. 21-2.
14. 'In a period of fierce competition between colonial powers, Indian teak, the most durable of shipbuilding timbers, saved England during the war with Napoleon and the later maritime expansion'. Gadgil and Guha, *This Fissured Land* . . . , p. 119.
15. K. Sivaramakrishnan, *Modern Forests: State Making and Environmental Change in Colonial Eastern India* (OUP, Delhi, 1999).
16. Sumit Guha, *Environment and Ethnicity in India, 1200-1991* (CUP, 1999); pp. 121, 129, 150, 163 and 'Conclusion'. D.E.U. Baker, *Colonialism in an Indian Hinterland: The Central Provinces, 1820-1920* (OUP, Delhi, 1993).
17. Nandini Sundar, *Subalterns and Sovereigns: An Anthropological History of Bastar, 1854-1996* (OUP, Delhi 1998).
18. See Sumit Guha, *Environment and Ethnicity* . . . , p. 2.
19. Nandini Sundar, *Subalterns and Sovereigns*, see the whole section 'Rebellious Pasts', especially Chaps. 3-5.
20. Ajay Skaria, *Hybrid Histories* . . . , pp. 40-4.
21. For some representative views and my observations on them, see B.B. Chaudhuri, 'Tribal Society in Transition: Eastern India, 1750-1920', in Mushirul Hasan and Narayani Gupta (eds.), *India's Colonial Encounter* (Manohar, 1993), pp. 99-102.
22. Ibid., pp. 102-5.

Reinterpreting Tribal Livelihood Systems: Underdevelopment and the Local Political Economy in Central India, 1800-1940

Archana Prasad

This essay attempts to challenge the ways in which historians, sociologists, anthropologists and activists have seen tribal livelihood systems in India. In particular it focuses on the resource rich region of eastern Central Provinces that now forms the state of Chhattisgarh in contemporary India. Colonial anthropologists and several post-colonial writers have called the Baiga and the Gonds, who lived in the area, its original and indigenous inhabitants. It was also believed that these livelihood systems were stable and isolated in character and had little connection with the outside world. Consequently, the needs of the people were limited and these societies could be characterized as 'prudent societies' as far as their resource use was concerned. It was argued that the tribals lived in harmony with their natural world because of these factors and it was the colonial annexation that destroyed the peaceful stability of these societies.[1] Scholars like Ramachandra Guha and others have argued that monopoly control over forests by the state; commercial forestry and timber trade for meeting colonial needs signified the modern scientific forestry project that was the vanguard of the colonial watershed. The exclusion of local rights and the consequent destruction of the local livelihood systems was the most dramatic consequence of these developments.

While this analysis may be true at a very general and simplistic level, this essay argues that the nature of tribal livelihood systems had always been determined by its relationship between the local, regional and national agrarian economy. However, the nature of this impact was qualitatively different in the colonial period. Evidence suggests that scientific forestry and monopoly control over forest lands were only symptomatic of the wider process of tribal destitution through the spread of industrial capitalism during the colonial period. Tribal societies were transformed into underdeveloped adjuncts of imperial capitalism by linking forests to global markets for forest produce. This led to the disruption of local production linkages essential for sustaining tribal survival, and limited the scope for the growth and expansion of local production linkages and resulted in a loss of tribal livelihood. Hence, the loss of livelihood was linked less to the nature of exclusion and more to the nature of the differential integration of the tribals into the world imperial and capitalist system. This in turn unleashed a spate of identity politics that signified the complete destruction of the tribal local economy and its incorporation into the global capitalist system.

This essay will be divided into four sections. The first section will look at the nature of tribal livelihood systems in the Gondwana in the late pre-colonial period. The nature of survival practices in the Maratha times will be studied through the documented oral traditions of the time when the Gonds regarded themselves essentially as sedentary permanent cultivators. The conditions of their marginalization will be studied in the context of the changing nature of the zamindari system in the region and the Anglo-Maratha conflict that led to the annexation of the Chhattisgarh and the formation of Central Provinces. In the context of this the second section describes the nature of the Gond and Baiga livelihood systems before the formation of the Forest Department. The third section of the paper will look at the relationship between the destruction of the productive capacities of the tribal economy with the colonial land and forest management systems. Different aspects and facets of the colonial agrarian policy are described in terms of their specific intervention into the Gond and Baiga livelihood systems. The basis of the unequal exchange between tribal and other societies was laid due to the processes described in this section. The last section of the paper will examine the nature of impact of the colonial break on the tribal society. In this section, we also see the emergence of the Baiga identity as a community of shifting cultivators and the relationship of the formation of such an identity with the threats to their survival. The section will point towards genesis of the identity politics that has informed these regions in the contemporary period and led to the formation of separate tribal states.

I. Migrations and Tribal Livelihood in the Gondwana

Anthropologists and historians of central India have universally referred to the Baiga and the Gonds 'original inhabitants' and the kings of the Gondwana. In the light of this assertion it is pertinent to examine who exactly the Baigas and Gonds were and where they came from. Though there is no mention of the Baiga in early records, the origin of the Gonds as recorded by Jenkins leads us to conjecture that the history of the Gondwana influenced both Baiga and Gond livelihood patterns. Early colonial accounts of the conflict between Deogarh and Maratha armies suggest that the Raghojee's invasion and conquest of Deogarh resulted in the organization and settlement of Gond zamindaris inside the forests. The Gonds' sense of movements and migrations as a result of military conflicts are highlighted in the fact that Jenkins' informant refers to the displacement of Gonds after the Gowalees slaughtered them in the 'Great War'. Thereafter he notes that the Gonds went and settled in Chanda and in other parts of the Vidarbha region.[2] Gond history as recorded by Jenkins states that,

> Originally they {the Gonds} were in the country. Afterwards they went below the Ghats to Chanda . . . there were considerable number of Gonds in this country but very few below the Ghats. When Raghojee Bhonsle took possession of the Gadee of Deogarh the inhabitants both from the top and below the Ghats went as far as Wurda and now very few Gonds of the Koetoor cast remain of the Ghats.[3]

It is amply clear from the oral evidence that the Ghats and the highlands were hardly the first place that a large section of the Gonds had inhabited. In fact Jenkins points to two types of Gond migration: one before the Maratha occupation and the other after the conquest of Gondwana by Raghojee Bhonsle. In either case, the mid-eighteenth century habitations of the Gonds were migratory in character. Their patterns of movement usually depended upon the nature of political conflict in their territories.

Gond memory is also accompanied by recall of the days when the Gonds were wealthy and produced sufficient to survive. These days are traced to the period of the ruler of Bukht Bulund, the father of Chand Sultan, who ruled Deogarh in the seventeenth century. As Jenkins' informant tells him,

> Formerly in this country almost everyone used to bring up cows and buffaloes and all the inhabitants possessed great wealth—kodo, kutki, mandya were in abundance—in the country and none of them were without gold and silver . . . in this state the country continued till the reign of Bukht Bulund.[4]

Jenkins source is probably referring to the loss of livelihood in the Maratha period when Bukht Bulund's son, Chand Sultan was in the service of Raghojee Bhonsle. There is a clear indication that the Gonds of Deogarh were farmers who grew small millets on tracts that had a good productivity till the 1740s after which they were displaced.[5] Jenkins notes that Raghojee left no stone unturned to secure his dominions by replacing petty Gond and Gowalee families 'who were opposed to regular Government and replacing them with more industrious inhabitants'.[6] By giving non-Gond people rights of permanent cultivation in these lands Raghojee ensured that the Gonds were not able to resettle on or use the tracts seasonally for cultivation and were forced to depend primarily on the forest and uncultivated lands at the periphery of the forest. Thus, in the late pre-colonial period Jenkins' informer noted that the Gonds *dhurm* was

> That they should live in the jungle, cut trees and level the ground and make cultivation and through the labour of their own hands provide themselves with their provisions. This is their principal Dhurm.[7]

This passage is important because it reveals some important facets of Gond life in the period after Raghojee displaced them. It shows that the survival strategies of the Gonds involved living in the jungle and doing semi-permanent cultivation with the plough. This form of cultivation appears in later colonial records as *burra* cultivation. Though there is no mention of the existence of hunting and gathering here, in another instance Jenkins' Gond informant asserts that food gathering and hunting were important supplements. Nevertheless, the lack of attention to gathering in this period may indicate that it was not an important part of the Gond existence in the pre-Raghojee period. Rather the Gonds (a generic term used for all tribals of the region in that period) saw themselves as marginalized permanent cultivators by the turn of eighteenth century. It is during this period that dependence on forests increased considerably. However, it is also important to note that the Maratha political system also allowed for a decentralized system of land control and access to and control over land and forests remained with local institutions to a large extent and allowed at least some local productive forces to survive. Given the above analysis about the transformation of the Gond livelihood system it is pertinent to examine the conditions which allowed Gonds to retain some amount of control over their own resources even while their process of marginalization had gained pace after the invasion of Raghojee.

The root cause of the beginning of the settlement of the Gonds into the forests can be traced in the changing nature of the zamindaris in the eighteenth century itself. In the seventeenth century, the Gondwana formed a part of the Mughal Empire. The decentralized form of resource use and control under the Mughals was significantly different from that under the Marathas. The raja of Garh Mandla, a principal chief of Gondwana under the Mughals enjoyed considerable autonomy over his territories. This autonomy was also a result of the distance between the empire and the chiefs and therefore Garh Mandla was ruled like an independent kingdom as long as they were loyal and paid tribute to the Mughal rulers. The subedar collected tributes and dues from the chieftains and left all the management functions to them. The rajas in turn granted some rent free tenures to the tribal people and ensured that their access to and mobility between the hills and the plains was protected. The raja allowed them to migrate seasonally and extract minor forest produce as long as they gave a portion of it to him as a symbol of their subordination. Interference in landholdings was minimal as neither the subedars nor the emperor showed interest in settling Hindu castes on tracts that were occupied by Gonds at the foothills of the Gondwana. This situation changed markedly with the beginning of Maratha rule in the early 1730s. In 1731 Raghojee became the ruler of the *suba* of Berar with the help of the raja of Deogarh. He received a tribute from the Gond raja of Rs. 10 lakhs a year as *chouth* and *sirdeshmukhee* as acknowledgement of his political superiority over the region. Both Chanda and Deogarh territories were subject to frequent attacks by his army after 1737.[8]

But even after the direct intervention of the Marathas, Jenkins' records show that Gonds were not tenants or people with land grants like the Brahmins and other castes. They were probably people who cultivated land at the pleasure of their chieftain, as is reflected in the piece of iron given to him every year. Access to land and forest was, thus, a result of a privilege granted in return for assistance whenever required by the ruler.[9] This assistance was, more often than not, in the nature of serving in the Gond raja's army or providing agricultural or household labour. Though the English records tell us little about the organization and the revenue settlements of Gond villages, there is sporadic evidence to suggest that these villages enjoyed some kind of autonomy as far as their use of land and forests were concerned. While zamindars did not restrict access to forest, they extracted certain dues from the tribal people in their territories including labour time and a portion of the extracted forest produce.[10] Gond peasants were also exempted from tax like *pandree* (toll tax collected from cattle carrying merchandise or grain) or *kosa baree* (collected on *tussar* worms) because of the services they rendered to the Gond raja.[11]

However, the relationship between the Gond zamindars and peasants were not conflict-free even if Gond peasants enjoyed certain concessions from their immediate rulers. The nature of the political system itself bore within the seeds of the conflict and the contradictions in the relationship between the peasant and the zamindars. The Gond zamindars' tendency to extract revenue and destroy the villages of ordinary villagers in order to escape the enemy was a common one in the wider Maratha-Gond conflict. The zamindaris themselves suffered a great deal in this wider conflict and this in turn led to the social, economic and political marginalization of the Gonds into the forest. The zamindars were repeatedly uprooted and driven into the forests after every military conquest and they in turn ravaged their subjects who tried to defend themselves against the warring parties. There are many instances to show that the zamindars looted and ravaged the villages of peasants and tribes. In the unequal relationship between the zamindars and their subjects, the biggest casualty was the regularity of the Gond peasants' residence in one place, thus having an adverse impact on agricultural activity. This led to a greater dependence of the peasantry on the forests for their survival. It also led to the increased agricultural activity in forested areas. However, even this level of survival was only possible when local access to land and forest resources was protected.

The mobility of the zamindars between plain and hill regions in the

Maratha period was in stark contrast to the permanent confinement of these zamindaris to the forest areas after the consolidation of territorial control under the early British regime. The British saw themselves as benefactors of the Gonds. They wanted to cure the Gonds of their habitual depredations, as they believed that these tendencies were borne out of the insecurity and the wild nature of Gond people. The cause of the insecurity was perceived to be conflict with the Marathas. The British saw themselves as the saviours of the Gonds from the mismanagement of the Marathas that was defined in terms of the prevailing chaos that was created by the constant movement of people in the wake of the continuous conflicts between the Maratha rulers and their Gond subjects. The British campaign in the Gondwana was thus designed to maintain the tranquillity of the ceded districts of the Nagpur state from the 'ravages of the wild and hitherto lawless tribes who plundered fertile tracts with impunity'.[12] As a part of this campaign the Gond chiefs incurred the wrath of the British Resident in Nagpur by making the escape of Appa Sahib possible after the Maratha War of 1817. The resultant victory of the British forces led to the surrender of Maratha and Gond chieftains by September 1819. Those who did not surrender migrated further into the hills carrying on their free-booting on the western side of the Narmada and in the territories of the nawab of Bhopal.[13] From then onwards the British government began settling their estates and determining the conditions of the surrender. The main aim of the settlements was to establish the loyalty of the zamindars and control the rich land, forest and mineral resources of the region.[14]

The British recognized that the question of the settlement of zamindaris was complex. Through these settlements, the British wanted to maximize revenue and minimize political instability and dissent. Many colonists wanted to bypass the zamindars' right to assess their own lands and create a system of settlements that prevailed in other British ruled territories. Jenkins, the British Resident at Nagpur, laid down the principles by which these zamindaris should be assessed. The first guideline suggested that *obarree* or quit rent should only be paid on 'hilly tracts that are unproductive'. In the next guideline he stated that the rest of the lands should be assessed on the basis of a five yearly *malguzari* tenure in which cultivating zamindars would be given preference. Since hardly any Gond zamindars cultivated land this principle was bound to benefit non-Gonds. Finally

Jenkins stated that hereditary patels and *goantias* (including zamindaris) should be incorporated into the colonial system of agrarian settlements in all areas.[15] The significant point about these settlements was that many landholding zamindari tenures were converted into revenue collecting *malguzari* tenures, especially in the plains areas.[16] Extending these principles to Garh Mandla, Colonel Molony stated that the continued tenure would only be possible in places which were either inaccessible or where the farmers could carry out improvements on their lands. All land of 'superior quality should be under the system of periodic assessments. According to this criteria there was no need to show any leniency to farmers in Garh Mandla, as the productivity of its land would be high if monsoons were good.'[17]

The conclusion of Gond settlements along these lines thus led to the diminishing of the territorial power of the Gond zamindars. Zamindari tenures were confined to hill and forest tracts and the forests formed an important part of their valuable resources. This development severely curtailed the movement patterns of the Gond peasants who were earlier exempt from taxes and enjoyed free mobility in zamindari tracts. Because of this, it is possible to surmise that the flexibility of the Gond subsistence economy was severely curtailed. Statements by Richard Jenkins' Gond informants show us that the process of the marginalization of Gond rights began in the late eighteenth century, when Raghojee Bhonsle expanded his territorial control over Gondwana. The Gond settlements of the 1820s intensified the pace of this process and diminished the fiscal, political and military power of the Gond zamindars in a more permanent and severe way. The abolition of zamindari tenures on fertile cultivable lands and the marginalization of zamindari rights into inaccessible forests affected the livelihood of the tribal people. The resultant occupational structure was a precarious combination of shifting cultivation, hunting and gathering activities.

II. Tribal Livelihoods and the Local Economy

The British action resulted in the annexation of the Nagpur state in 1854. Between 1818-54 the territories of the Nagpur raja were administered under a residency and after the annexation, the first seven years of direct British rule were spent in bringing a majority of the resources, land and people under their direct control. It was only after this that the Central Provinces were formed as an administrative unit of the British Empire in 1861. Some zamindari rights were also granted in lands and areas that were in the inaccessible highlands. Coincidentally, this is also the first time when the British records mention the Baigas as one of the most prominent tribal ethnic groups.[18] It is therefore pertinent to ask who these Baiga were and how they constituted themselves as a separate community.

Baiga subsistence depended on a combination of gathering, hunting and swidden cultivation practices for their survival in the early nineteenth century. Of these, swidden cultivation was considered the cornerstone of Baiga livelihood. To this end the debate over the ecological and economic viability of *bewar*, a Baigani term for shifting cultivation, dominated discussions on Baiga survival. However, a closer inspection of their survival strategies reveals that swidden cultivation or *bewar* cannot be seen in isolation from other survival practices. In Baiga livelihood gathering, hunting and *bewar* were bound together in a precarious balance by a fragile seasonal rhythm, whose success was intrinsically dependent on the success of the *bewar* cycle. Unlike other forms of permanent cultivation, *bewar* was only cultivated on the same field for three consecutive years, the cropping patterns and agricultural cycle differing from one year to the next. Scholarship on the Baigas and other forest communities has not taken these hunting and gathering activities as an intrinsic part of seasonal survival. Instead they have considered them as auxiliary or supplementary activities.[19] Yet if we analyse the linkages between these three activities it is possible to surmise that the shifting cultivation system is difficult to sustain without the existence of hunting and gathering.

The swiddening practices of the Baigas were similar to those of the Khonds of Orissa. Therefore, a comparative analysis of the relationship of the survival practices of the two communities will help to make the connections between *bewar* and different subsistence activities used to make this connection more explicit. In *Chait* (March-April) women went to clearings to pick castor and cut stalks that were already ripened. The women gathered fruits like sago, tamarind and mushrooms. Baiga women gathered only roots and *mahua* (*Madhuca indica*) seeds at this time.[20] Both Khond and Baiga women did the gathering, men cut large trees and prepared the *bewar* fields. Men of both communities also went on their ritual hunt that began at full moon. In this context, Elwin notes that Baiga were known as the best hunters amongst all forest communities in the Central Provinces. They used bamboo bows, arrows, and spears. The spears were made of iron and sometimes consisted of a bamboo shaft with an iron head. The bows and arrows were generally made of the *dhamin* (*Grewia elastica*) wood. Traps were only used to protect the fields.[21]

In *Baisakh* (April-May), the firing of the forest took place, and the Khond women gathered unburned wood to burn. Men continued to hunt, but only in nearby forests. The seeds were sown in *Jeth* (May-June), and hunting continued. From *Asadh* (June-July) to *Bhadon* (August-September), men worked in the fields. In *Kuar* (September-October), the first beans were ripened and in *Kartik* (October-November), small millets or *kutki* became ripe. In *Aghan* (November-December) every crop was ready and in *Pus* (December-January) winnowing took place. *Pus* was also the time for dances and marriages. In *Magh* (January-February), shifts were made to new *bewars* and hunting-gathering was the main source of survival. This was the agricultural cycle of the first year. In the second year, there was more time for hunting because fewer crops had to be sown and harvested, but since there was enough food, the people lived on the *bewars*. It was only in the third year that the diet had to be supplemented with the forest products and game from hunting.[22]

The Khond case shows the dynamic nature of seasonal survival in non-sedentary cultivation practices. It indicates that the dependence of the community on forms of survival other than cultivation was maximized in the first year. *Bewar* was prepared for the first time in that year and cultivators were occupied in clearing the fields. Therefore dependence on hunting and gathering activities was almost total. The interaction between *bewar* and other seasonal activities was minimized in the second year. This was primarily because the community could depend on the crop from the first year. During this period, hunting and gathering was used for both ritual purposes and for supplementing the diet unless there was a crop failure in the first year. In the third year, a combination of both cultivation and hunting-gathering were the basis of survival. The communities depended upon the crops of the second year and fruits, roots, etc., from the forest. Thus, the three-year cycle of survival was intrinsically dependent on the success of the crop of the first year. If it failed, the precarious balance between these three modes of production would be disturbed.

It is therefore evident that the importance of *bewar* in the Baiga livelihood system varied from year to year and from one season to the next. This was true even as *bewar* formed the fulcrum of the seasonal cycle and enjoyed a dialectical relationship with hunting and gathering. Without them, *bewar* itself would become unsustainable. If *bewar* were to be practised on its own, then either its fallow cycles would become shorter, or the Baigas would be compelled to move to other places to cut more fields. Further, hunting and gathering provided important inputs to diet by providing meat, roots and food during times when there was either a drought or when the crops failed. Therefore, it was important for them to coexist with shifting cultivation practices in order to ensure that the needs of Baiga life could be met. It is because of this link between *bewar* and other practices that a minimum level of forest diversity had to be maintained in order to ensure that the population dependent upon *bewar* did not face a food shortage in case of the failure of the *bewar* crop.

Apart from this, another reason for the maintenance of the diversity was the need to maintain the fallow cycles of *bewar*. The main aim of these fallow cycles was the need for the regeneration of forest undergrowth in order to replenish the nutrients of the soil. If the undergrowth failed to appear, it was impossible for the cultivator to do *bewar* cultivation on the field.[23] What swiddening did not require was the growth of large timber trees. In fact there is sufficient evidence to show that the cultivation cycle often interfered with the natural regeneration of *sal* trees that had been felled in the preparation of the *bewar*. Early surveys showed that good *sal* was found on the undulating slopes of the Maikals. Where the soil drainage was good, the best *sal* would grow also on low lying plains.[24] But many of these areas did not have *bewar*. Most of the *bewar* fields were found on moderate slopes where thick undergrowth and mixed forests prevailed over mono-*sal* forests.[25] In these areas *bewar* coexisted with the following species:

Sal grew with drought resistant trees like *saj*, *salai*, and *aoula*. It was also surrounded by a variety of grasses like mosaic, *khel*—the principal grazing grass—and *sukra* in the western Maikals.[26] John Nisbet, an official of the Indian Forest Service, enumerated the advantages of a forest where *sal* grew with other trees. He said that such forests had a greater density, and better retention of the soil. The forester argued that the natural reproduction of these forests was easier

TABLE 1: FOREST TREES AND GRASSES IN THE MAIKALS

English (Latin) names	Local name	Uses
Sal (*Shorea robusta*)	*Surye* or *Rinjal*	Seeds are used for food and resin and also for medicines
Black Eyne (*Terminalia Tomentosa*)	*Saj*	Used for breeding lac and its resin is sold to local craftsmen for dyes
Butea frondosa	*Palas* and *Dhak*	Useful for dyes and tans local craftsmen and its flowers are used by local people for ornaments
Bosewellia thurifera	*Salai*	Drought resistant tree used by Agarias to make charcoal
Scheilchera trijuga	*Kosum*	Oil is generated from its bark and the lac insect prefers this tree
Embilica officinalios	*Aoula*	Its fruit is eaten during famine
Grewia elasica	*Dhamin*	Used for making bows, arrows and traps
Madhuca indica	*Mahua*	Used for liquor and food
Dichanthium caricosum.A˙	*Khel*	Principal grazing grass
	Sukra	Medicinal grass

SOURCES: From *Mandla DG* 1909; and Forsyth 1889.

than that of forest with a single species. Last, such forests were less prone to natural disaster than pure forests.[27] They thus helped to neutralize the onslaughts of *bewar* on *sal* forests, and also helped to protect the tribal communities like the Baigas from scarcities and famines. So far I have argued that Gond and Baiga subsistence was intimately linked to their interrelated and multiple occupational structures. I have also hinted that this was only possible in a forest with diverse forest species which made seasonal gathering possible as well as provided a market for some of this produce. This is so because forest gathering was undertaken, not merely as a supplement to other occupations, but as a seasonal livelihood strategy. It required the Baigas and the Gonds to have local linkages with peasants and artisans who either visited the forests or used forest produce.

Let us now consider the nature of some of these networks and the ways in which they brought about the integration of the Baiga and Gonds into the wider local political economy. An examination of gathering activities in forests shows that Baiga survival was intimately connected to their links with peasants, graziers and craftsmen who were important users of the forests. The Baigas believed that the forest had everything they needed for sustenance. Since their primary source of survival was the forests, the Gonds, Pankas and even lower caste Hindus considered them superior to the other tribes because of their important ritual role. Because of this the Baiga saw themselves as kings of the forests. This image was firmly grounded in the material basis of Baiga survival. The Baiga used the forests for gathering food, *sal* seeds and other minor forest produce. Both women and men undertook gathering activities. Women gathered roots, fruit, leaves, *mahua*, and *sal* seeds. Root and fruit gathering were important sources of food in periods when *bewar* crop was ripening. Men also gathered roots in times of drought and famine. According to Russell and Hiralal, a Baiga could collect a full basket of roots in a single day.[28] *Mahua* and *sal* seeds were collected by women and used as food or for brewing *tadi* (country liquor). Apart from this, the collection of leaves was useful for household purposes and bamboo was collected to make baskets and mats that were sold in the markets.

Gathering was not a gender specific task. Both men and women did the gathering for the local agrarian market. This brought them in contact with people who did not live in the forests but depended in one way or another on them. The nature of the contact was, however, quite gendered in at least some respects. Men sold more lucrative produce like honey, lac sticks, silk cocoons, and tans to local craftsmen. In contrast the women sold flowers, leaves and seeds of trees like the *sal* as well as fuel wood to peasants. These were presumably less lucrative than the produce sold by men and this could have led to male control over any surplus in the household income. The domination of male members in controlling the most profitable ways of survival was also revealed in the relationships that informed the cultivation system. The male members of the community controlled the technology and the means of production. For example the women were not allowed to even touch the axe. The father was the head of the family, but the son could own his own axe after marriage. Once he got married the father gifted an axe to his son. The gift of the axe was an important gesture that recognized the son as a co-equal producer. At the same time, it also created a hierarchy between households—the domination of the patriarch continued till the son attained that position himself. The male child's access to the axe, bows, and arrows gave him power. The same was not true of the women.[29]

The Baiga patriarch reinforced his position vis-à-vis the women in their relationship with the peasants. Even as the Baigas considered cultivation by the plough a sin, and most peasant castes as socially inferior, the structure of the regional political economy ensured that peasants were wealthier and had more social influence than the Baigas. Apart from excluding the women from the decision-making process, the self-image of Baiga men as protectors of mother earth had little effect on their social and economic position within the wider economy. They continued to eke out their existence from the precarious combination of swiddening and gathering. But by refusing to work as wage labourers, the men forced the women to do the bulk of the work and perform the tasks of transplantation and harvesting, especially during times of scarcity when *bewar* crops had failed. Men only hired their labour out for one specific task: the uprooting of rice seedlings before transplantation.[30] In this way the Baiga men refused to be integrated into the local agrarian economy as oppressed servants.

The relationship of the Baiga with the Agaria iron smelters and grain traders was more clearly indicative of the differential integration of the Baiga men and women into the local society. In Mandla, the main job of the Agarias was to repair old sickles and axes and to supply new ones. For this they received an annual fee of grain from the villagers, called *jewar*. Then they got food grain in the month of *Asad*, known as *bij phutti*, and finally they received grains at the time of the harvest of both spring and autumn crops, known as *kainhari*. The Agaria rates depended upon the nature of the soil. In the zamindaris of Bilaspur all trade took place through the medium of iron. Instances of exchange between the Maria and the Agaria were common in Bastar and Chanda.[31] The presence of the Agaria linked the Baiga to the wider agrarian economy. In times of famines, the Agarias played an important part in sustaining the Baiga economy. They sold them food, oil and salt for everyday needs. Despite this linkage, the Baigas considered the Agarias outcastes. They lived in settlements outside the village when they visited it. They were only called inside the village when they had to repair implements or perform some rites on a newly born child with red-hot iron.[32] The women of the two communities hardly met each other and contact was kept largely through the men. Since the Agarias were mobile smelters and traders, they would have carried information about a particular region and locality.

The Baiga economic system was thus one in which men and women not only performed a different job but were also linked to the local society in a different way. Though both men and women found themselves at the lower end of the local relations of production as reflected in their relationship with the peasant society. Peasants, forest dwellers shared forest lands on the fringes of ryotwari villages, and specialist herdsmen like the Ahirs, Gowaries and the Banjaras. The Ahirs kept and grazed cattle in the forests for individual peasant families. Their intimate relationships with the Gonds and others living in the forest were a result of these daily excursions. Russell notes that in Chanda, the Ahirs had developed intimate links with the Gonds. They not only employed Gonds to graze cattle but also admitted them within the Ahir caste. In fact, the origins of the Gowarie community of Chanda were traced to inter-marriage between Gonds and Ahirs. In Mandla, villagers thought that Ahirs were not part of the Hindu community as they lived with the Gonds.[33]

The relationships of the Banjaras with the peasant and forest communities followed a different pattern. An early article on the migratory people of India described the Banjaras as grain merchants who carried their wares on pack oxen. They traversed the impracticable country to get supplies that they took into districts where scarcity prevailed, and later moved to large army camps where they supplied grain. Thus, they established a considerably important status in the political economy of the Central Provinces.[34] The Banjaras made their living from the sale of salt and oil to forest dwellers and the transportation of goods. Ethnographers have given instances of the Charans and the Banjaras carrying loads for the Rajput and the Maratha armies.[35] The Banjara seasonal movement spanned a large territory. Their ability to negotiate thick forests and 'rugged' terrain made them indispensable to the rulers who relied on them to carry messages and arms. The British often condemned them as robbers and bandits due to this. They carried salt and oil for the forest dwellers and were essential for their counterparts in the forests. Their utility to the peasant economy was minimal. Though they sold a few milk products to the peasants, they never grazed peasant cattle. This was done by more 'trusted' graziers and residents of the village, like the Ahirs and Gowaries, who even though low castes, were certainly included in the system exchanges within the village community.

But the linkages of the tribal economy were not limited to inter-village contact. The forests saw a convergence of different occupational interests that supported the artisan and peasant economies. The mobility of the artisans between towns, villages and forests ensured the development of wider linkages of forests with other economies. The lac jewellery and toy makers were good examples of this. The craftsmen who worked with lac did not necessarily live inside the forest because upper caste peasants and craftsmen regarded lac propagation with certain superstitious repugnance. Therefore these craftsmen bought stick lac from the Gonds, Korkus and Baigas.[36] The lac seed swarmed twice a year, in December and June. Labour for its collection was available in June-July for the *Baisakh* crop and in October-November for the *Kartik* crop. Thereafter lac was taken to the markets where the forest dwellers sold it to the craftsmen. The main lac market near Mandla was in Seoni where lac was in great demand.[37] After collection, stick lac picked of the wood, and then placed the encrusted twigs and barks in long cotton bags. These bags were heated and the lacquer gum squeezed from the sticks and mixed with clay and other materials to make toys and bangles. Since the best business for the Lakheras was in the season of festivals, their income was seasonal and any transaction of the forest communities to propagate lac must have needed to take that into account. While their main markets were in small towns like Mandla and Seoni, they also sold bangles in the villages. The presence of lac artisans in the suburbs of the towns and the mobility of the Korkus, Gonds and the Baigas to sell the lac created wide-ranging linkages between the urban, agricultural and forest societies. The tribes found themselves at the lower end of these networks. They had to walk long distances to get to the markets and had to sell their lac at prices on which the artisan agreed. This is so because the Lakhera or artisans who moulded lac, were superior to the Baigas in the social hierarchy.[38]

Similar linkages between urban centres and forests were evident in the case of the dyers. The main dyers of the province were the Koshtis, Chippas, Nillars, Rangaris and Rangrez. They dyed in traditional colours, mainly reds and yellow got from the roots of trees like *al*, *kusum* and *rohun*. Resin was extracted by tree tapping, and in some cases, by pounding roots. Forsyth does not record whether resin or gum was extracted by the dyer or bought from the forest dweller. Rangrez and Rangaris wove silk-bordered cloths and dyed their own thread with lac or *palas* flowers. Since they abhorred manual labour, it is likely that they bought stick lac, and *palas* flowers from forest people. Apart from this, the *tussar* silk industry and the silk weavers were dependent on the collection of silk cocoons by Gonds. Silk weavers, like Koshtis bought cocoons from the forest communities and boiled them, and wound the tread on reels. Dyers used myrabalans flowers to get blue or black, while red dye was extracted from lac. Other cotton weavers like Chippas, Nillars and the Rangrez also bought flowers of a similar kind to make dyes for cloths. The dyers did not necessarily develop close social relations with the forest communities. Those like the Koshtis lived in towns and enjoyed a good income. The class difference gave them an economic advantage over the forest communities. Conversely, the hesitance of these artisans to go into the forest to collect their own resins also made them dependent on mobile forest dwellers.[39] The multiple networks described above resulted in defining the framework of the local economy, with the Gonds and Baigas playing an important role in the life of the graziers, artisans and peasants depending upon the nature and location of the forest. But in all cases, the tribes of the Central Provinces found themselves at the lower end of the scale. Their access to productive resources and land became restricted with the changing nature of the state in the eighteenth and nineteenth centuries.

III. Tribal Survival and the Colonial Intervention

The settlements of land, coupled with the monopoly state control over forests and minerals were an important part of the strategy to integrate tribal labour and knowledge into the larger system of colonial extraction and global capitalism. Three developments in the colonial period had an adverse impact on the productive links between the tribal and the larger agrarian economy. The first, as mentioned repeatedly in this paper, was the land settlements of the early and mid-nineteenth century. The second significant development was the enactment of the first comprehensive forest laws in 1878 and the consequent attempt to create enclaves of tribal concentration that would serve as labour camps for the forest department. The formation of the Baiga Chak and forest villages in the 1890s was a prime example of this development. The third most important development from the point of view of tribal survival was the penetration of industrial capitalism in the forested areas from the late nineteenth century onwards when non-timber forest produce became a valuable commodity in the world market. The penetration of the managing agency into the forested areas signified the integration of the tribal economy into the global capitalist system. Thus, the access of the tribal society to its productive assets was squeezed from both forests and agricultural lands.

The Forest Department of the Provinces was formed in 1865 and in the same year the government also declared that all forests, except for forests on which private zamindari and *malguzari* rights had been settled, would be the property of the state. Thereafter the surveys of forested areas laid the basis of their future classification and also determined the rights of the local people in forests. In the Act of 1878 that followed these surveys a rigid classification of forests was made the basis of forest conservancy. In the process the Baigas and Gonds lost their mobility and free access to the forests. The classification of the period between 1878 and 1886 shows that almost three-fourths of the forested tracts were closed to local use. Simultaneously the British also embarked on their civilizing mission by trying to introduce the Baigas to plough cultivation. Both these developments had an adverse impact on the survival of Baigas whose livelihood system hinged on the successful completion on the *bewar* cycle. This prompted the Baigas to flee to the zamindari forests in order to eke out their livelihoods.[40]

However, one notable feature of the above mentioned phenomenon was the constraints that the migration from government forests placed on the nature of the scientific forestry project itself. Foresters of nineteenth century noted that the fleeing of the Baigas from *khalsa* areas had created a labour shortage for forest works. It was thus imperative that the Baigas be provided some inducement and means of livelihood in order that they return to the *khalsa* areas. It is for this reason that we see a significant shift in the colonial agrarian policy towards the Baigas in the late 1880s. The first step towards squeezing out the Baigas from the zamindari areas were taken by the enactment of the rules for the administration of zamindari and other private forests. Three rules were formulated to give the government wider control of the forests: (a) the District Commissioner was to advise the Chief Commissioner on whether zamindari forests were to constitute a separate *mahal* for the purposes of settlements; (b) the settlement was to be fixed for a period of three years; and (c) there was to be a stipulation for the making of working plans for these forests and that the forests were to be earmarked for reservation at the discretion of the District Commissioner. In the past it was observed that the zamindars had often bypassed the government in affairs concerning forest management. For this reason they were to sign an agreement to comply with these rules. By this measure the government would be able to appropriate the largest share of the zamindari profits and keep an account of the zamindari income. This in turn would check the practice of leasing through 'illegal' contracts. Apart from this, *bewar* was also banned in forested areas of the zamindaris.[41]

This measure coupled with the administration of government forests effectively banned all cultivation in any forested areas, but did not go without protest. In the first instance, the zamindars of Bhandara, Chanda and Balaghat began protesting the rules.[42] However, gradually Baiga villagers themselves started petitioning the government for permission to practice *bewar* in the forests. The Commissioner of Nagpur, commenting on this petition said that,

> If bewar is completely withdrawn the Baigas will do one of the two things—either they will cut bewar at all costs or they will emigrate to other areas like Kawardha or elsewhere.[43]

The officials realized that the Baigas were unlikely to settle down to sedentary cultivation. Therefore, the objective of the policy was shifted from inducing the Baigas to use the plough, to restricting *bewar* cultivation within a limited area.

Thus, the District Commissioner of Mandla, Mr Miller, favoured the establishment of a separate Baiga settlement. Within this settlement the Baigas were to be induced to take up the plough. But if this failed, at least their *bewar* cutting could be controlled. In July 1878 Miller reported that 200 Baiga families were willing to settle under such a scheme.[44] In 1890 two villages and four *tolas* or settlements of houses were chosen to form the 'Baiga reserve'. Originally comprising 362 families, these villages were in the Dindori and Karanjiya forest reserves. Tracts were to be allotted only to Baigas and were to be available for further colonization by them. Thus began the process of the formation of the Baiga Chak. The Chak consisted of 22,858 acres of land that was reserved for *bewar* cutting for the Baigas. It consisted of five villages and nine *tolas* (cluster of houses). These villages were

Silpiri and Dhurukota in Karanjiya range and Dhaba, Ajgar and Rajnisarai in Dindori range. The *tolas* were Silpiri, Tantar, Dhurukota and Kandwani in Karanjiya range and Dhaba, Ajgar, Rajnisarai, Lamotha and Jhamul in Dindori range. They were put under the jurisdiction of the District Collector and the Tahsildar.[45]

There is little information about how many *bewars* were granted during this period. But there is a record of the Baiga protest at the insufficient allocation of land for their *bewars* and the displacement that they suffered due to the formation of the Chak. Dharmu Baiga of Mouza Oudhar wrote to the tehsildar about the 60 Baiga families that were to be removed to Dhurukota:

> I am residing in Mouza Oudhar and I am cutting bewar in this village for the past 3 years. Sixty families of the Baigas inhabit these villages and they all live upon Dhya cultivation. We have now been ordered to leave this village and remove to Dhurukota. . . . It has very little jungle and most of it was cut down by old Baiga settlers; and consequently it cannot support other Baigas of different village. . . . We therefore pray that we may be allowed bewar cutting in adjoining villages.[46]

The 'policy of colonization' (within the Chak) was important for the success of this plan. The contraction of the land available for *bewars* and increasing population density were the only factors that were likely to push the Baigas to take to the plough. In the beginning the reservation of forests was used as an excuse to move people. By 1891-2 it was reported that in Ramgarh tahsil only 80 families living outside the reserve had taken to permanent cultivation willingly. For those Baigas who did not either want to go to the Chak or take up the plough the government decided that four villages would be set up at Karadih, Udhar, Pandpur and Duldul for the supply of labour to the forest department. These villages were not to be solely Baiga villages but Gonds and Agarias were also to be settled here.[47] In 1895 Gonds and Agarias were allowed to take up land inside the Chak, furthering the 'policy of acculturation' that was first started in Balaghat.

While the making of the Chak was the result of a process of negotiation between the Baigas and the British, it gradually became clear that its creation would hardly solve the problem of Baiga survival. Despite the formation of the Chak it was noted that the *bewar* crop was systematically failing because of shorter fallow cycles and that the Baigas were forced to depend more and more on forest and other seasonal labour for their livelihood. Many of them also began to live in forest villages that began to serve as labour camps for the forest department. The first forest village regulations were issued in 1890. Under these laws forest villages could be established within the limits of any 'reserved' forest with the prior consent of the Conservator. The District Commissioner and the Divisional Forest Officer (DFO) would decide their location. Forest villages were to be designed solely for the permanent supply of labour and were not to be made with the intention of extension of cultivation. Lastly forest villages were to be made up of those communities that were 'habituated to the extraction of forest produce'.[48]

These measures became a part of the scheme to conserve forests and control tribal cultivation in forests. They also established administrative control over labour supply and the exploitation of the tribals by contractors and outsiders and through this process facilitated the monopoly control of the colonial state over forest labour. The forest department had a claim to the labour of all adult villagers: no employment could be taken up without the consent of the DFO. The DFO was solely responsible for the administration of these tracts. Each household was to be allowed a patch for cultivation at fixed rates per plough; free grazing was to be allowed for a limited number of cattle. The forest dwellers were also to be allowed free thorns and wood to repair their houses, and would have the first claim to forest work for which they were to get fair wages. As far as the village organization was concerned the patel or the headman was to be appointed from amongst the villagers and was to enforce the collection of rents and cesses and see to the sanitation of the area. All advances for cultivation were to be forwarded by the forest department and this department was also to be responsible for health and education.[49] The first results of the establishment of forest villages were evident in 1895. Between 1892 and 1895 the department established 16 forest villages in Mandla, 8 in Jabalpur, 4 in Narsingpur, 26 in Betul and 23 in Nimar. 98 villages were established in Balaghat, along with 29 in Chanda, 7 in Chhindwara and 7 in Seoni by 1897. There were 454 forest villages in the Province by the turn of the century. R.H.E. Thompson, the Conservator of Forests even held that the forest dwellers were reaping 'considerable advantages from labour opportunities'.[50] Thus, the Forest Department had arbitrary control over these villages. They regulated the forest community's relationships with contractors (if the forest was leased) and determined what kind of production and labour relations would be developed amongst forest dwellers.

However, the relations of production and exchange were not only determined by local factors but also global and national trends. The late half of the nineteenth century saw the advent of industrial capitalism in forested areas. Several minor forest produce were important raw materials for the leather, handloom, gunpowder, paper and other industries in this period. They were also important from the point of view of the artisanal economy that depended on tribal forest produce collectors for raw materials like gums, resins, flowers seeds, roots, etc. In this way the tribal economy was linked to both the rural and the urban society. Interventions in these linkages began in the mid-nineteenth century with the *nistar* and grazing laws and continued till the first quarter of the twentieth century. This was especially evident in the case of the changing demand patterns for minor forest produce from the 1880s onwards. The rise in the world demand for minor forest products led to the entrance of European managing agencies in trade of forest products. It also led to some experimentation in the substitution of these products by chemical tans and dyes and this led to a substantial change within some artisanal industries. At the same time some artisans also felt the adverse impact of the changing nature of the minor forest produce trade as shown in the case of the lac industry. These developments severed the production relations between the artisanal and tribal economies and then led to their incorporation within the colonial division of labour.

The changing nature of the tribal economy and its incorporation in colonial industrial capitalism can be best elucidated through the example of the propagation of lac in the Central Provinces. The first evidence of rising international demand for minor forest produce was seen in the rising prices of lac that had many industrial uses. It was an insulating agent chiefly used as varnish and could be used as stiffening material for felt and straw. Further it was useful as a preservant in coating wood or metal and could also be utilized as an adhesive material. Due to these functions lac was in great demand in America, England and Germany and the exported lac was often converted into shellac in these countries in the nineteenth century.[51] The initial attempts to modify the ways in which lac was propagated and new varieties of lac introduced, failed in the Central Provinces. The government noted that since the province was not capable of yielding lac of real value it was not worthwhile for the government to take up lac cultivation on its own. Therefore, it was considered better if private agencies and contractors were given the right to propagate lac.[52] The firms and managing agencies had to incorporate the local methods of lac propagation in their extraction of lac for export. In 1875 the government debated whether it should grant Messrs Jardine and Skinner, an international managing agency, a contract for lac propagation in the Loisinga and Barapahar forests of Sambalpur. The District Commissioner of Sambalpur, Bowie reasoned that,

> The propagation of lac is only carried on by Gonds, Binjiwars and other jungle tribes who are poor and always require advances to survive. While they propagate lac the government can only give advances if it has the lease of the jungle. By taking a royalty, the interests of the government and the firm will be kept identical.[53]

This official assertion of the compatibility of the Gond, official and industrial interests was one of the first steps towards the inclusion of Binjiwars into the world market. The royalty and advance were indicative of the presence of the European agencies in the forest economy. In the wake of such demand there were two courses open to the government. First, they could allocate leases according to the axioms of a competitive market. The highest bidder (for the payment of royalty) was to get the contract. While the cultivator had a right to sell his labour at a price he decided, in practice the lessee was the only purchaser of this labour. The lessee was to pay the labour for lac propagation and fix the wage rate according to the number of maunds produced. Thereafter the lessee acquired an ownership right over produce. According to the logic of this system, the government royalty would remain static, while the lessee's profits rose or declined according to the fluctuations of the market. Alternatively, the government would plan to receive a minimum royalty (to be fixed at 25 per cent) of the income earned through the sale of produce by the lessee. Under this system the government would fix the wages of the lac cultivator as well as the price at which the lessee could sell lac in the market. The maximum price at which the lessee was allowed to sell lac in the market was Rs. 10 per maund or Rs. 30 per bullock load. By fixing the price thus, the government attempted to narrow the gap between the profit of the leaseholder and the price paid to the lac propagator.[54]

By 1919 the colonial government claimed that the methods used by the forest communities were inefficient for mass propagation. R.S. Troup contended that the methods of local lac propagation were inadequate in at least two ways: the expense involved in searching for the lac bearing shoots and large quantities of lac are lost due to the time taken to collect lac.[55] The Forest Research Institute, where Troup worked, carried out lac experiments to see the extent to which these disadvantages could be minimized. But as Troup pointed the experiments led to no conclusive results in the techniques for propagating lac significantly. Troup held that the ambivalent results were a result due to the fact that the experiments were carried out in the lower Shivaliks, a region distinctly unsuitable for lac cultivation. However, he suggested some conditions under which lac cultivation could be carried out more effectively. The annual pollarding of lac trees, the growth of trees in an open position and the need for thinning trees more regularly were identified as some of the desirable steps to be undertaken.[56] While these experiments were being carried out the Chief Commissioner proposed the introduction of machinery in lac cultivation. He felt that by introducing technological innovations in forests the production of lac could increase and the production process made efficient. This would save the effort of watching the lac throughout the season, and confine their labour to the collecting process. He refuted the proposition that the employment opportunities of the Binjiwars and Gonds would be affected if this happened and instead stated that the measure would help tribal lac collectors to strike a better bargain with the representatives of managing agencies. In order to maximize production and assess the value of lac several government sponsored experiments were carried out in the early twentieth century. Several techniques were tried to improve the quality of the seeds, minimize labour required and reduce the injury to trees. But the reported failure of all efforts (according to the special lac officer), proved that the methods used by forest dwellers and the peasants were more effective.[57] This prompted European managers to incorporate local techniques for lac propagation within its system of collection and production. The efforts to mechanize the propagation of lac were confined to rearing the cocoon and not to the collection of produce. An attempt was made to increase the productivity of lac cocoons by improving the propagation methods of lac. However, it is important to note that these attempts and measures were not implemented on a mass scale and lac continued to be harvested twice a year through the methods that had been used by the tribals in the past. One of the main reasons for this was the need for the continuous and rapid supply of lac and shellac to the European industry in the inter-war period. It also saw a significant difference in the nature of trade. While in the nineteenth century significant amounts of shellac was being produced outside the country, in the pre- and inter-war period some shellac producing units came up in urban areas on the hinterland. For example, the Divisional Forest Officer of Bilaspur Division noted that the button and shellac company had been established in Champa by a European firm to reduce the charges of the middleman and save on freight carriage to Europe.[58] By the 1940s there were 35 shell lac factories in Chhattisgarh that produced 16 per cent of the lac in the entire country.[59]

The influx of European capital in lac provided the forest communities with seasonal employment in the forest areas. The penetration of European capital in the propagation of lac helped to create labour opportunities for the Bhumias and Gonds who started working for managing agencies like Jardine and Skinner.[60] The adverse first impact of this was seen on the inter-linkages between the artisan and the forest dwellers. The forest dweller started supplying lac to the agents of the European firm instead of the artisan. The leasing system created monopolies of managing firms over forests and labour, pushing out smaller lac artisans from the market. It also put tribal lac collectors at the mercy of European capitalist firms whose main interest lay in using cheap labour to propagate lac and export it. In this context the main failure of the lac experiments lay in the inability to develop the conditions of their successful conclusion without incurring a substantial cost. For this reason the effort to introduce new technologies in lac production need to be grounded in the willingness of the managing agencies to incur some infrastructural cost to induce their success. Thus, the very logic of Imperial capitalism forced the colonial regime to rely on tribal techniques and labour for the fulfilment of their objectives. This in turn elucidated the nature of integration of tribal labour in the world capitalist system. Thus the modification of the tribal livelihood system in the process of its integration within the system of colonial exploitation was disastrous and led to the long term underdevelopment of the tribal economy.

IV. Underdevelopment and Tribal Identity Politics

Thus far we have traced the changes within the tribal livelihood systems from the late eighteenth to the early twentieth century. We have also seen that the main feature of these changes was the complete destruction of the productive capacity of the tribal people and their conversion into cheap providers of labour and raw materials. Herein lies the genesis of unequal exchanges that later characterized the movements for tribal self-rule in the region. However, the irony of the matter is that an aggressive tribal identity got strengthened even as the material basis of its existence started to disappear and pre-capitalist forms got gradually incorporated into the colonial and capitalist mode of production. By the turn of the century the differential impact of colonial measures could be seen on the Gond and Baiga societies. But in all cases tribal people were the poorest of poor and the colonial state was compelled to enact welfare measures for them so that they could continue to provide a hard working and able-bodied cheap labour force.

Agriculture was the first sphere where the marginalization of tribal people was seen. The nature of excision induced people to take up tenures in relatively low producing tracts and hoped that tribal people would take up some of these tenures in the 1890s. However, the Settlement Officer of Mandla, Mr Bell held that the protection of Gond *malguzari* rights was essential if favourable conditions were to be created for increasing the number of tribal landholdings.[61] He argued that the Gond *malguzars* had lost their rights and lands because of constant expropriation by non-tribals. He substantiated this by stating that theoretically speaking, the Gonds should have held majority of the lands in the district. However, they were only in control of 116 out of 1364 villages as most of the estates were transferred to non-aboriginals due to debts.[62] This had a direct impact on the status of the aboriginal tenants in the district. By the time of the settlement of 1904-10, around half of the recorded Baiga population was classed as poor who were dependent on their labour for their livelihood. Only seven Baigas were considered rich tenants, or those who were free from debt. The rest of the Baigas had incurred some form of debt or the other. As far as the Gond tenants were concerned, 80 per cent of them were classed as peasants with some form of debt or the other. One-third were categorized as very poor whereas only 20 per cent of the Gond peasants were free from debts.[63] Such analysis by the settlement officers of various districts prompted the enactment of the Central Provinces Tenancy Act, 1916 that was mainly designed to protect tribal holdings. But by the end of the first quarter of the century it was amply clear that though the settlements of the 1920s had steadily shown an increase in the tribal holdings of the Provinces, the average size of these holdings was declining more and more. This made the Baigas depend more and more on labour as they could not pursue any other occupations because they were 'educationally and politically backward'.[64]

The situation on the labour front was also not satisfactory and signified the further impoverishment of tribal people. As late as the late 1930s and early 1940s the main source of subsistence in forest villages remained forest labour. In 1940 the Governor Wylie wrote that cash payments were not enough to meet their requirements and he called upon the Forest Department to have a more humane approach towards the problems of forest labour. He questioned the scale of wages paid to labour for felling and carting and demanded an early report on the subject. He also spoke of the problem of piece-work when he said that Baigas were made to labour on roads till they were physically in a most unsatisfactory shape. Thus, he concluded that the conditions under which these Baigas worked affected their health. Lastly, the Baigas were exploited by the forest department, as there was a departmental extraction of 'illegal and forced labour' during harvest and sowing time. The Forest Department made the labourers work more than 8 hours a day without paying them extra money. According to Wylie this was equivalent to the practice of forced labour. Wylie demanded careful consideration of all these points if the forest department wanted the active cooperation of the forest dwellers.[65]

In this context of extreme poverty and penury, the self-perception of Baiga people underwent a rapid and dramatic change. In the nineteenth century the Baigas emphasized their role as the kings of the forests, whereas in the twentieth century *bewar* was emphasized to strengthen primordial Baiga authority. The decrease in cultivation by the Baigas also led to their dependence on gathering and labour for their survival. This in turn gave the Baiga women a greater role in the fulfilment of Baiga needs. At the same time Baiga men no longer held the weapon that they used to marginalize the women: their monopoly over the control of the means of production. The latent conflicts over power and control between two segments of the community were witnessed in the articulation of an aggressive Baiga identity vis-à-vis other communities. It increased the power of the Baiga patriarchs, the cornerstone of this articulation was still *bewar* cultivation and the social and cultural power of the Baigas over other people. The cutting of each *bewar* is an event in the exercise of power against hostile forces. It was an act in emotional sustenance, an act of 'joy' and the movement associated with it, the movement of life. This was seen in the personal life of many Baigas that were recorded by people like Elwin, who played an important part fixing this image of the Baiga in the minds of his contemporaries.[66] All history was epitomized in the 'self' constituted around *bewar*. The events of Baiga history, could be divided into two periods: the time of 'joy', i.e. the period of unrestricted *bewar* and the period of 'no joy' when the restriction on *bewar* cutting started. The first period is packed with accounts of the great deeds of the Baigas who were meant to be experts at hunting tigers and as medicine men. The second period has descriptions of the evil deeds of witches signifying the advent of a bad omen. This period had instances of disease and famines as their major hallmarks. Though none of the Baiga remembers dates of events, the coming of the British was an important event for all of them and was seen as the end of prosperity. In most of the personal accounts the fulfilment of needs became a problem with the coming of British. The British were described as robbers. The restrictions on *bewars* were described as nothing more than a subjective problem of the *sahib*. Yet this attitude expressed the general attitude of the Baigas.[67]

By the 1930s, Baiga behaviour become aggressively anti-British in defence of their identity that was signified by *bewar*. What happened to the Baiga, whom Ward had once described as 'soft spoken and ever obliging'? That the Baigas were defiant towards the British is evident from the following statement 'Send us if you will to jail, for now the jungle is closed all the world is a jail to us. Take back your bullocks for we don't need them.'[68] The refusal to accept the ban over *bewar* was justified by a self-perception that their entire social being was signified in *bewar*. The origin myth recorded by Elwin in the 1930s clearly showed this:

The Baigas were established in the practice of bewar by Bhagwan himself, who when he called the tribes of the world to make a king first chose the Baiga. But Nanga Baiga begged that the Gond his brother be made king instead. Bhagwan was pleased, and as a mark of favour, took Nanga Baiga by the hand and placed him on his throne by his side. He granted his prayer to make the Gond King, but gave him a greater blessing.

'All kingdoms of the world', he said, 'may fall to pieces but, he who is made of the earth and is Bhumiraja, lord of the earth shall never forsake it. You will make your living from the earth. You will dig roots and eat them. You will cut wood and carry it on your shoulders. Your wife will pick leaves and sell them. You must not tear the breast of your Mother the Earth with the plough like the Gond and the Hindu. You will cut down trees and burn them and sow the seeds in their ashes. But never become rich, for if you did you would forsake the earth and there will be no one to guard it and keep its nail in place.'[69]

The Baiga identity was getting constituted vis-à-vis the British, the other communities and by stressing their individuality the Baigas remained together as one community, despite occupational changes at least in the first decade of the twentieth century. Thus, the reproduction of social ideology and the changes in oral traditions did not always reflect the changes in material reality.

However, a section of the Baigas had come to grips with their situation, and understood the power structure of the colonial regime. These Baigas probably lived in the Khalsa areas where the control over *bewar* was stricter than in zamindari areas. Where earlier the Baigas would lie about their bullocks,[70] now at least some of them started lamenting its existence. With the bullock came the demand for the removal of restrictions on *bewar* and reduction of taxes, for more employment and more food. The differentiation between Baigas also increased. For example, it was held in 1937 that there was one Baiga within the Chak who was lending money to others.[71] Within the Baiga identity a class identity developed. There was now a 'rich' Baiga like the moneylenders, a Baiga cultivator with the plough, and a 'poor' Baiga labourer. Though both of them still believed in Nanga Baiga, the Baiga identity implied different things to different people. Even though this despair reflected the changes in material reality, the Baigas still saw plough cultivation and the British as 'evil and sinful'. Nanga Baiga and his life stopped being the blueprint of daily life, rather it became a symbol of solidarity and a device to understand the wide-ranging agrarian changes. Some Baigas do not conform to traditional community norms to cope with wide-ranging agrarian changes. This in turn led to the declining authority of village elders and priests. In contrast to this, the tendency of the Baigas to focus the articulation of their community identity around *bewar* cultivation reaffirmed the power structures that regulated individual behaviour within the community. Between these two opposing tendencies stood the colonial state and the complexity of social relations with in which the Baigas placed themselves as a 'superior race and the guardians of mother earth'. The reproduction of the social ideology and material reality was simultaneous. Nanga Baiga influenced the course of Baiga history in the colonial era and also on the changes in their material reality. Though the changes in material reality, however, did not always reflect the changes in the oral traditions, they did influence the ways in which the Baiga identity articulated itself. This identity concentrated on demanding the rights to cut *bewar* and did not adequately stress on the multi-occupational dimension of Baiga subsistence. Thus, while the identity of the Baigas as *bewar* cultivators was a unifying feature of the community in an emerging class differentiation, it also signified the extreme penury and marginalization that had set in because of the measures taken by the colonial government.

This form and expression of identity, however, underwent a rapid change with the emergence of a dominant ideology of tribal welfare in the 1940s. The underdevelopment of tribal society, brought with it the desire to adopt measures that would dress the wounds of the tribal people. The debate that the Act of 1935 fostered, considered special measures for the protection of tribal people and created excluded and partially excluded areas that attempted to isolate the forest dwellers from others.[72] But the most notable feature of this debate was the focus on what welfare services were to be provided to the people inhabiting these areas. The main issues discussed were education, health, cultural protection and some form of meagre livelihood support. By the mid-1940s, the isolation of tribals from the rest of the society had got institutionalized under the Policy of Exclusion and Partial Exclusion that was ratified by the legislature under the Act of 1935. This isolation resulted in the destitution of tribal people who were slowly becoming dependent on other agencies for meeting their daily needs.

The debate on how these services impacted upon tribal culture and life led to the consolidation of the 1930s debate into an ideological and political battle between the Christian missionaries and the broad alliance between the Congress and other Hindu nationalists and Elwin. The missionaries were the first people to practice the philosophy of tribal welfare in mid-nineteenth century. However by the 1940s, they faced stiff resistance from the Congress, Hindu Mahasabha and anthropologists like Verrier Elwin. The reports of the Congress Nationalists, Hindu Mahasabha and Verrier Elwin in the early 1940s suggested that Mandla was becoming a haven for proselytism by the Christian missionaries who were the earliest to arrive in these tribal areas. Elwin was the earliest to note the increasing activity of the missions. He noted that the missionary activities had increased markedly since 1941-2. He and Thakkar also complained that the District Commissioner of Mandla, a Musalmaan, was supporting the missions.[73] Thus, in one stroke, both A.V. Thakkar and Elwin created an opposition between Hinduism and other religions. The change in Elwin's attitude was remarkable. While earlier he had gone out of his way to argue for the protection of a distinctive tribal identity, now he began to argue that tribals formed a part of the broader framework of Hindu culture.

The results of the contest between the religious clerics of the Christian and Hindu Missions were seen immediately. In the same report the *pracharak* also stated that they had managed to reconvert some individual tribal families. An instance of mass reconversion and purification was also reported in 1945 when a Hindu Sangathana week was celebrated in the district. A conference was also held to establish a branch of the Hindu Mahasabha in Mandla district. The conference passed the following resolution,

> Gonds, Pankas, Baigas, Pradhans and other castes of the place from about 20-25 places attended the Hindu Conference. The assemblage amounted to about 5000-7000. In the conference a resolution to the effect that Hindus of the place will not render any help to any caste, creed or institution which puts the whole Hindu community at a loss and will increase the Hindu community after sacrificing the life and will resist anti-Hindu activities.[74]

These events led to processes that attempted the consolidation of the tribal identity as a Hindu identity and created a conflict between the activities of Verrier Elwin (working in tandem with the Arya

Dharam Sevak Sangh and Congress) and the missions who were the oldest reforming agency in the region. This conflict, as evident in Elwin's report of activities in 1944, emphasized that reconversion of tribals to Hinduism was a nationwide phenomenon by the mid-twentieth century and fuelled the need of non-missionary activists to lobby for the ban of missionary activities in partially excluded areas.[75] The anti-missionary campaign, with Elwin as its chief protagonist in Central Provinces, demanded that,

> Missionaries should be withdrawn from Partially Excluded areas, we insist that all education in these areas should be taken over by the Government. We demand that the Government should do twice as much as the missionaries have hitherto achieved. We have no interest in keeping these people backward. If they are to take their place as Kshatriyas in the Hindu social system, then they must be trained in the arts of liberal thinking and educated to courage and the traditions of honour.[76]

Written at the twilight of colonialism, this statement highlighted the fact that anti-missionary activists had started believing that assimilation into Hinduism was one of the only ways of tribal development. This view gained greater currency in the early post independence period. The constituent assembly debates of 1949 redefined the terms on which tribal areas were to be partially excluded in different states. The Fifth Schedule under which partially excluded areas were to be governed, were to be administered by the Governor with the help of the Tribal Advisory Council that was to consist of tribal representatives in considerable numbers. They were not to be excluded from the working of the Constitution except in respect of provisions concerning marriage, social customs or land distribution if the Governor deemed it necessary.[77] Mandla and all other partially excluded areas of the Central Provinces were included in the Fifth Schedule. The provisions of the Fifth Schedule were referred to a select committee. However, the debate that ensued showed clear signs of the colonial legacy. Most members who opposed the Fifth and Sixth Schedules blamed the missionaries for sowing the seeds for separatist tendencies and the demand for a separate tribal state.[78] And even though these Schedules were finally institutionalized, the events of the 1960s led to the banning of missionary activity in the Fifth Scheduled areas of Madhya Pradesh and facilitated the expansion of Hindu nationalist agencies in the later period.[79]

Conclusion

This paper has attempted to trace the genesis of the identity politics that plagues the central Indian tribal region and has led to the formation of separate tribal states. It shows that the root cause of these developments lay in the destruction of the productive capacities of the tribal people. This left them with little option to survive and they were forced to become providers of cheap labour and raw materials if they were to meet their meagre needs. This resulted in their integration into the larger system of colonial and capitalist exploitation that led to the complete penury of tribal people. Our analysis has also shown that tribal peasants were in an equally piteous position and were marginalized into lands with low productivity. This led to their indebtedness and they too were forced into seasonal forest and other labour. These developments led to the emergence of an aggressive tribal identity that was quite divorced from their material reality. The advent of a policy of tribal welfare has to be seen in this context. It has been argued in this essay that the tribals were left with no option but to depend upon the welfare services that were to be provided to them by the first half of the twentieth century. This dependence on welfarism for their survival also led to a politics of competitive communalism that had a far-reaching impact on the tribal polity in the later half of the twentieth century. Similarly the creation of scheduled areas and the formation of an aggressive and primordial tribal identity led to the demand for self rule that culminated in the formation of new tribal states in the year 2000. However, the issue of the restoration of the productive assets and capabilities of tribal economies, the root cause of the decline in tribal society remains unaddressed. Until this is done no amount of self rule can lead to the resurgence of the social and economic life of tribal central India.

NOTES

1. For this analysis see especially Ramachandra Guha and Madhav Gadgil, *This Fissured Land: An Ecological History of India* (OUP, Delhi), 1992.
2. MSS EURO F.34, India Office Library (hereafter IOL), p. 138.
3. Ibid., p. 320.
4. Ibid., pp. 330-1.

5. For more details about Gond self-perceptions and the details of their marginalization as permanent cultivators, see Archana Prasad, 'Military Conflicts and Forests in the Gondwana', *Environment and History* (Cambridge, 1999).
6. MSS. EURO. F.34, IOL, p. 88.
7. Ibid., p. 321.
8. Ibid., pp. 13-17.
9. Ibid., p. 321.
10. For this point, see W.V. Grigson, *The Maria Gonds of Bastar* (London, 1936), pp. 14, 18, 141-3. Also see Glasfurd, *Report on the Territories of Bastar*, 1876, n.p.
11. Col. Agnew, *Report on the Administration of the Subah of Chhattisgarh* (Nagpur, 1830), pp. 14-29.
12. Board Collection, F/4/755/20541 of 1820 (IOL, London), pp. 136-7.
13. Ibid., p.142.
14. Ibid., pp. 158-60.
15. Board Collection, F4/796/21428 of 1824 (IOL, London), pp. 1-11.
16. Ibid., pp. 66-8.
17. Ibid., pp. 191-2, 295.
18. An exception to this is Stephen Hislop, *Papers Regarding the Tribes of Central Provinces*, Scottish Mission Nagpur, 1840. However, here too the Baigas are described as a larger sub-sect of the Gonds. There is also just one reference to the Baiga as a priest in his pre-colonial study. The first detailed study is that of Colonel Ward in 1866-70 when the first settlement operations were started in the Mandla District. The next most prominent survey is that by Bloomfield in the 1890s when he surveyed the Chak. Thereafter, Verrier Elwin published his seminal work *Baiga* in 1939. Verrier Elwin, *The Baiga* (OUP, London, 1939).
19. J. Forsyth, *The Highlands of Central India: Notes on their Forests and Wild Tribes, Natural History and Sportly* (London, 1889); Russell and Hiralal, *Tribes and Castes of the Central Provinces of India* (London, 1910), Vol. 2; Mahesh Rangarajan, *Fencing the Forest: Conservation and Ecological Change in India's Central Provinces, 1860-1919* (OUP, Delhi, 1996).
20. Verrier Elwin Papers, No. 165, 'Notes on the Khonds', Nehru Memorial Museum & Library, New Delhi (hereafter Khond Notes), p. 89.
21. Elwin, *The Baiga*, p .87.
22. Khond Notes, p. 109.
23. Archana Prasad, 'The Political Ecology of Swidden Cultivation', *Tools and Tillage*, Vol. II, No. 4 (Denmark, 1995).
24. R.C. Ostmarton, 'Sal and its regeneration', *Indian Forester*, Vol. 54, 1928, pp. 569-70.
25. Archana Prasad, 'Political Ecology of Swidden Cultivation'.
26. *Gazetteer of Mandla District*, 1909 (hereafter Mandla DG, 1909).
27. John Nisbet, 'On Mixed Forests and Their Advantages over Pure Forests', *Indian Forester* (1894), p. 5.
28. Russell and Hiralal, *Tribes and Castes of Central Provinces*, Vol. 2, p. 91.
29. Elwin, *The Baiga*, pp. 78-9, 273.
30. Russell and Hiralal, *Tribes and Castes in Central Provinces*, Vol. 2, p. 9.
31. *Mandla DG*, 1909, pp. 179-80; Verrier Elwin, *The Agaria* (OUP, Delhi, 1942).
32. Russell and Hiralal, *Tribes and Castes of Central Provinces*, Vol. 3, p. 122.
33. Ibid., Vol. 2, p. 24.

34. Edward Balfour, 'On migratory tribes of the natives in Central India', *Journal of the Asiatic Society of Bengal* (Calcutta, 1844), pp. 2-3.
35. N.F. Cumberledge, *Monograph on the Banjara Clan* (Nagpur, 1882); Col. Mackenzie, 'Notes on the Banjaras', *Berar Census Report* (1881); Russell and Hiralal, *Tribes and Castes of Central Provinces*, Vol. 2, pp. 163-92.
36. Russell and Hiralal, *Tribes and Castes of Central Provinces*, Vol. 3, p. 122.
37. Central Provinces Secretariat Records, Nagpur, Forest Department Compilation 50(6), 1875 (hereafter Lac 1875), pp. 85-6, and Central Provinces Secretariat Records, Nagpur, Forest Department Case File No. 22, Prog. 'B', April 1919 (hereafter Lac 1919), p. 3.
38. Russell and Hiralal, *Tribes and Castes of Central Provinces*, Vol. 3, p. 106.
39. Ibid., Vols. 2 and 3.
40. For greater details of this phenomenon see Archana Prasad, 'The Baiga: Survival Strategies and Local Economy in Colonial Central Provinces', *Studies in History*, December 1998, Delhi.
41. Government of India Records (NAI), Agriculture and Revenue Department (Forests) (hereafter ARF), Prog. 'A', Nos. 1-4 of July 1889, pp. 1-2, and Central Provinces Secretariat Records (hereafter CPSR), Forest Department, Compilation No. 158 of 1888-90, p. 61.
42. For details of the zamindars petitions, see Government of India Records (NAI), Agriculture and Revenue Department (Revenue) (hereafter ARR), Prog. 'B', No. 3 of February 1892.
43. CPSR Forest Department, Compilation No. 178 of 1881, p. 7.
44. Ibid.
45. Mandla District Report (hereafter MDR), Mandla Collectorate. Forests File No. 37 of 1937, p. 105.
46. CPSR Forest Department, Compilation No. 178 of 1890-2, pp. 11-12.
47. Ibid., pp. 22-3.
48. ARF Prog 'A', Nos. 16-17 of May 1891, p. 1.
49. Ibid., pp. 2-3.
50. *Report of Forest Administration, Central Provinces* (hereafter FRCP), 1895-1900 (combined volume), pp. 36-8.
51. Lac 1919, p. 51.
52. Lac 1875, pp. 32-3.
53. Lac 1875, p. 8.
54. Ibid., pp. 26-30.
55. R.S. Troup, 'Experiments in the pollarding odd Butea Frondosa for lac cultivation', *Indian Forester* (May 1919), p. 225.
56. Ibid., pp. 227-8. Troup and his team carried out experiments over ten years. They divided the forested tracts with trees of different girth and ages into strips and applied the lac worm at different times. At each time they recorded the amount of lac that they got from the tree. In this way they determined the ideal conditions for the harvest of lac.
57. Madhya Pradesh Secretariat Records, Bhopal (hereafter MPSR), Forest Department, File No. 114, September 1920.
58. James Best, Messrs. Becker & Co's Lac Factory at Champa in Bilaspur District, Central Provinces, *Indian Forester* (1912), p. 514.
59. *Provincial Industries Committee Report, 1946* (Government Press, Nagpur, 1947), p. 67.

60. These opportunities were important for their survival in the wake of restrictions over forest use especially after the reservation of forests in 1878. Other authors like Rangarajan, *Fencing the Forests*, and Ramachandra Guha, 'Forestry in British and Post-British India', *Economic and Political Weekly*, 1985, have described this process.
61. *Final Settlement Report of Mandla District, 1904-1910* (Nagpur, 1910), p. 40.
62. Ibid., p. 42.
63. Ibid., p. 43.
64. Grigson, *Aboriginal Problem in the Central Provinces* (OUP, London), 1944, p. 83.
65. Forest Department, Eastern Circle Conservators Office, Jabalpur (hereafter FDECCO), Case File No. XVI-M of 1940/121, pp. 1-4. Wylie quoted Symington to emphasize this perspective, 'The State has a clear right to extract from all its forests all the profits that can *justly* be derived from them but justice demands that forest labourers should be treated somewhat better than slaves ...' (see p. 2). The rest of this file has a debate in which Grigson and the Commissioner of Jabalpur disagreed with Wylie, but the statistics showed otherwise.
66. Elwin, *Baiga*, pp. 132-40, has 13 biographies of Baiga men and women reflecting this.
67. These conclusions are taken out of a reading of the above mentioned biographies.
68. Ibid., p. 51.
69. Elwin, *The Baiga*, pp. 106-7.
70. Ibid, p. 151, The Baigas often ran away if they did not want to use the plough.
71. MDR Forests, File No. 37 of 1937, p. 53.
72. See Nandini Sundar, *Subalterns and Sovereigns: Ethno-history of Bastar* (OUP, Delhi), 1997 and Savyasaachi, *Tribal Forest Dwellers and Self-Rule: The Constituent Assembly Debates on Fifth and Sixth Schedules* (Indian Social Institute, Delhi, 1998).
73. Verrier Elwin, 'Report by Bhumijan Sewa Mandal on Activities of Christian Missionaries in Mandla', 1, 1944. The manuscript of this document is in Bhulabhai Desai Papers (NMML), and Correspondence with Verrier Elwin of 12 October 1944.
74. All India Hindu Mahasabha Papers (NMML), 1945, File No. M-11, pp. 153-4.
75. Purshottamdas Thakurdas Papers (NMML), File 337 of 1949, Report of Verrier Elwin on Anti-Missionary Activities, dated 17 December 1946.
76. 'Missionaries and Aboriginals', 7, typescript in Bhulabhai Desai Papers, NMML, Correspondence with Verrier Elwin of 12 October 1944.
77. Provisions of the Fifth Schedule are reproduced in Savyasachi, *Tribal Forest Dwellers and Self-Rule: The Constituent Assembly Debates on Fifth and Sixth Schedules* (Indian Social Institute, Delhi, 1998), pp. 54-8.
78. Ibid. Debates on the north-east particularly show this.
79. For a longer version of this argument, see Archana Prasad, 'Conversions and Tribal Identity in the Central Provinces', *Alpajan*, 2000, No. 1.

Tribes, Soldiers, and Social Change in Early Nineteenth Century Malwa

Amar Farooqui

When the British established their supremacy over the Malwa region in the first quarter of the nineteenth century they justified their political and military intervention, as they usually did, in terms of the chaotic conditions then prevailing in the area and the inability of the indigenous states to cope with the situation. John Malcolm, the principal colonial official in Malwa after the third Anglo-Maratha War, wrote a lengthy memoir on the subject. The dominant themes of this well-known Memoir of Central India, including Malwa and Adjoining Provinces are the conflicts among the Malwa states, especially the prolonged struggle between the Sindia and Holkar states to establish their hegemony, and the Pindari problem. The Pindaris were singled out by Malcolm as the most disruptive factor in the region. One might mention in this context that when Lord Hastings embarked upon a fresh round of colonial expansion in western and central India he regarded the destruction of the Pindaris as a matter of great urgency. What is grandly referred to as the third Anglo-Maratha war actually began as a campaign against the Pindaris. Hastings had repeatedly asked for permission to launch a war against the Pindaris and even though a conditional sanction was given he was told that 'the authorities at home are unwilling to incur the risk of a general war for the uncertain purpose of extirpating altogether the Pindaris'.[1] Eventually Hastings was able to convince his superiors that the annihilation of the Pindaris was the key to establishing British supremacy over Malwa.

It needs to be underlined that when the British entered Malwa they encountered a society that was undergoing profound changes. In a sense the Pindaris exemplified these changes. The Pindaris must be placed against the backdrop of a continuous historical process of social differentiation in the extensive central and western Indian tribal belt. This process was linked to state formation, empire building and the search for military recruits. The entire Madhya Pradesh-Rajasthan/Gujarat border, and the Vindhya-Satpura region, still has a substantial tribal population. Throughout the medieval period the communities inhabiting this area were going through a phase of transition to class society. In the later medieval period the pace of this transition had been accelerated by Mughal expansion, the drive of the Dakhani saltanats, and Maratha empire building. It is now recognized that the inhabitants of the central Indian hills and forests did not lead a completely isolated existence. They were suppliers of forest produce so that there was some exchange between them and settled agriculturists; strategically and commercially important routes passed through the area inhabited by them; and they could provide human resources to patrol the hill and jungle terrain they knew so well. While discussing the evolving ethnicities of Bhils and Gonds, the two major tribal groups of the central Indian forest, Sumit Guha points out that the tribal dominated lands of this region 'were no tranquil backwaters, isolated from the turbulent politics of the peninsula'. 'Nor indeed were they, he adds, inhabited by relic populations obstinately preserving their archaic lifeways. They were in fact deeply integrated into the political economy of medieval India.'[2]

One important trend set in motion by changes taking place in this frontier region was the emergence of warriors in tribal societies, which was linked by Surajit Sinha to state formation, the articulation of new identities, and Rajputization.[3] The tribal belt would henceforth provide recruits for armies, while petty 'predatory chiefs' would add to the pool of those whose services could be enlisted by contractors for recruits.[4] Dirk Kolff's study of the phenomenon of recruitment of troopers by middlemen in medieval India, *Naukar, Rajput and Sepoy*, contains a useful discussion on Rajputization and the transition from a tribal to a warrior culture based on differentiation and specialization. Kolff refers to a broader development, which includes the Pathans, whereby transition to a social formation marked by the presence of a professional warrior class led to dissolving former identities and taking over readymade new ones defined either as Rajput or Pathan. 'This may be a heretical thing to say', Kolff argues, 'but I suggest that, according to the ways of the north Indian military labour market, in the pre-Mughal period, "Afghan" as well as "Rajput" were soldiers' identities rather than ethnic or genealogical denotations. It was merely to register membership of the war-band they had decided to join that until quite late on Indian soldiers were known by such names.'[5] One would like to suggest that whatever the term 'Pindari' might have originally meant, by the second half of the eighteenth century it had come to denote an intermediate identity in Malwa and the middle Narmada. This was yet another option available to the new warriors for whom it defined an identity which could, given the right opportunity, become a sort of stepping stone to more reputed identities.

Kolff's work helps us to comprehend what was happening on the periphery of peasant societies in central India in the later medieval period. There was a connection between military service, state formation, and newly acquired identities (Rajput, Pathan, Pindari). It was an indication of a society in a state of flux that social groups were not 'closed'; identities were not yet inflexible. Mughal, Maratha and Rajput states absorbed the soldiers provided by the tribal belt. Pindari chiefs became, in effect, contractors or sub-contractors looking after the interests of their respective groups and making no permanent commitments. Guha argues that 'Maratha incursions in north-western Khandesh could hardly have been possible without at least the acquiescence of the local Bhils. Their support was also enlisted by the Mughal commanders'.[6] The support of this soldiery had to be constantly renegotiated. Whenever possible or necessary they could go in for small-time plunder or extorting money for ensuring the safety of vital routes passing through their settlements.

The short-lived Pindari phenomenon offers some insights into these developments—despite the paucity of information on the Pindaris. We hardly know anything in their own terms about their past, their traditions, and what distinguished them as a group.

Highly prejudiced servants of the East India Company are our main source of information about the Pindaris. This information is supplemented by references to them in Maratha records. It was during the second Anglo-Maratha war that colonial officials began taking notice of the Pindaris and lost interest in them soon after 1818. The views of Malcolm were influential in moulding the colonial perception of the Pindaris. Malcolm first began scrutinizing the Pindaris when he was at the Sindia darbar during the first half of 1804.[7] Later, in his *Memoir* Malcolm presented a detailed and connected narrative of the

Pindaris.[8] This was to remain, and still remains, the standard account of the Pindaris. Most of what has been written about the Pindaris can ultimately traced, apart from Malcolm's *Memoir*, to contemporary reports prepared by James Tod (1811), Richard Jenkins (1812) and George Sydenham (1814), and the voluminous correspondence *Papers Respecting the Pindarry and Mahratta Wars*.[9]

The official colonial view of the Pindaris focuses exclusively on their plundering activities, completely missing out both on the disruption which colonial intervention caused, and the twin processes of social differentiation and state formation that tribal people in parts of western and central India were undergoing during the eighteenth and nineteenth centuries. They were originally one of the numerous people living on the fringe of settled society in Malwa. Their integration began when many of them found a place in Maratha military organization. By the end of the eighteenth century they had become a regular part of Maratha campaigns north of the Narmada.[10]

Some of the hypotheses about the origins of the Pindaris are so vague as to render meaningless any discussion on this problem. In one attempt to unravel the mystery of their origins the term Pindari is linked to a variety of intoxicating drink to which the Pindaris are supposed to have been addicted;[11] in another attempt the name is linked to the word *pendha* or bundle of straw suggesting that Pindaris were originally foragers.[12]

More relevant is the reference to the tribal core of the Pindaris in R.V. Russell's *Tribes and Castes of the Central Provinces of India*. According to Russell the Pindaris had descended from the Gonds and Bhils of central India.[3] There was certainly an older Pindari tribal stratum which had a strong connection with the Narmada Valley and the Vindhyas. The *Settlement Report*, for Hoshangabad, a district in which there had been a number of Pindari possessions, observed that their ancestors 'were generally Gonds, Kurkus, Bhils etc.'. [i4] Right till 1818 the main Pindari chiefs retained holdings in the Narmada Valley and the Vindhyan forests, especially in the stretch between south eastern Dewas and the Hoshangabad-Narsimhapur border.[15]

Parts of Hoshangabad (district) remained closely identified with the Pindaris (Handia; Harda; Timurni; Makrai, etc.). Besides, some of the adjoining areas of Dewas—Nemawar, Satwas, Kannod, Kantaphor, Bagli—were Pindari bases. In Narsimhapur (district) the two most important Pindari chiefs, Chitu and Karim Khan, had holdings at

Barha and Paloha. Chitu had lodged his family at a village close to Satwas.[16] He had received Satwas as a grant from Daulat Rao Sindia, and had built a fort there. Four guns were reportedly placed at the fort. Satwas (district Dewas, Madhya Pradesh), lies south of the Vindhyas and commands the road to Nemawar. Nemawar (district Dewas) is located strategically on the northern bank of the Narmada at a point where the river is fordable. Handia (district Hoshangabad, Madhya Pradesh) is situated on the opposite bank. Nemawar, which was part of Sindia's territories, had also been granted to Chitu who established a cantonment at the place.[17] South of the Narmada, Timurni, Makrai, Harda, Barha and Paloha were among the territories held by the Pindari chiefs. It is worth noting that in the Narmada Valley most of their possessions were concentrated in the tract between Handia and Narsimhapur, i.e. east of Nimar and west of Jabalpur.

In the rather remote and inhospitable Narsimhapur area both Chitu and Karim Khan had accepted grants from Sindia. These appear to have had almost a symbolic significance. The two villages given to them (Barha to Chitu and Paloha to Karim Khan) could not have yielded a large revenue nor were they strategically important. They are situated close to each other in Gadarwara tahsil in the western portion of district Narsimhapur, not far from the southern bank of the Narmada. The Shakkar river, a tributary of the Narmada, runs along the west of Paloha. Barha lies south of Paloha. Chitu had built a fort at Barha.[18] By itself the Narsimhapur *area* was of little consequence at this point of time. Until about the last quarter of the eighteenth century it was unsettled and sparsely populated. Narsimhapur town, originally known as village Gadariya Kheda (settlement of shepherds), began to develop only during the 1780s.[19]

Chitu and Karim Khan did not have any particular personal attachment to the Narsimhapur area, since they themselves came from Mewat and Malwa respectively, but they were keen to maintain a presence at this gateway to the Gond homeland.[20] This would have allowed them to procure fresh recruits for their raiding expeditions, most of which increasingly took place in northern Malwa. As a matter of fact the Pindari leaders retained outposts in the Narsimhapur-Hoshangabad-Dewas stretch of the Narmada Valley, partly to have some safe refuge in case of an emergency, but mainly to enlist followers from among tribal communities of the central Indian forest

(Gonds, Bhils, etc.). The lifestyles of these communities had undergone a metamorphosis due to various historical processes. The resulting 'social disintegration' led to the formation of tribal bands 'in the hills and woodlands' which engaged in plunder to support themselves.[21] These groups had a crucial role to play in the Pindari scheme of things as suppliers of soldiers. Recruitment would have taken place during or immediately after monsoon when the Pindaris congregated on the banks of the Narmada to recoup their resources. They normally desisted from undertaking any expeditions till the rains were over.[22]

The Pindaris were not a closed, homogeneous group. In the words of Sydenham, 'they now form a heterogeneous mass containing followers of almost every caste of Hindoos, and every sect of Mahomedans'. Similarly, the commission which inquired into an alleged Pindari raid on Guntur and some other parts of Andhra recorded that the raiders 'were composed of all tribes, from the . . . Bramin and Rajepoot, to the Paria . . . '.[23]

Whatever little evidence we have on how the Pindaris interacted with each other indicates that members of Pindari bands did not share a primordial or permanent attachment to one another. The term Pindari does not suggest a membership composed of an entire ethnic or language group. Though several Pindari bands might have drawn together periodically, as for instance near the Narmada during the rainy season, there were no traditions that compelled them to gather regularly or permanently as a single unit. The major bands were in a constant state of flux as smaller groups departed, families veered off on their own for various periods and captives married into the community.

In the years preceding the third Anglo-Maratha War there were two outstanding Pindari leaders: Karim Khan and Chitu. Of the two, Karim Khan was politically more important having been on the verge of creating an independent state. Tod remarked in 1811 that Karim Khan 'had attained a degree of established power, before unknown to these plundering unsettled Hordes'. He added that Karim possessed 'a pretty extensive, compact fertile and well cultivated territory, containing several strongholds well garrisoned, and the appearance of a regular system of authority'.[24]

Karim Khan's territories were mainly located in eastern Malwa, especially on the outskirts of the Bhopal state,[25] which was almost on the verge of extinction after 1805. Bhopal, which had risen to prominence in the first quarter of the eighteenth century under Dost Mohammad Khan, was the main Pathan stronghold in Malwa.[26] Bhopal and the adjacent eastern Malwa parganas became the main target of the Pindaris. Karim Khan had established his principal cantonment at Berasia near Bhopal.[27] Sindia and Holkar preferred to confine the Pindaris to this area, at some distance from the core of their territory in western Malwa. For example, the tribute due to Sindia and Holkar from the frontier Rajgarh and Narsinghgarh principalities in north-east Malwa was formally divided between Karim Khan and Chitu who were left to deal with this difficult area.[28] For a long time the eastern Malwa parganas had been administered directly from Pune, but with the collapse of the Peshwa's authority they had been encroached upon by Sindia, Holkar as well as by numerous other contenders.

Significantly, the moment Karim Khan attempted to graduate to the status of an independent ruler, Sindia cut him to size. Karim Khan was held captive by Daulat Rao Sindia from 1806 to 1811. The Sindia-Karim Khan relationship reveals the contradictions inherent in the political crisis which preceded the third Anglo-Maratha War. Under the great Mahadji, the Sindia state had initiated a military modernization programme in the last quarter of the eighteenth century. Benoit de Boigne was one of the main pillars of this programme. De Boigne's departure in 1796, the defection in 1803 of General Perron who had succeeded de Boigne, and Sindia's losses in the second Anglo-Maratha War had been major setbacks to the modernization programme.[29]

After 1805 the Pindaris were looked upon as useful and relatively inexpensive military allies. There was a marked increase in the number of Pindaris employed as soldiers by the Maratha states of Malwa. Jenkins estimated that their number had doubled between 1804 and 1809, from about 13,000 to roughly 25,000.[30] At the same time he differentiated between 'real' Pindaris of the pre-1805 period and the large number of new recruits, many of whom were discharged soldiers of Indian states which had been forced to scale down their military strength or disband their armies altogether after the second Anglo-Maratha War.[31] In 1811 Sindia entered into a specific agreement with Karim Khan 'that no other persons but such as were real Pindaris should be entertained in the durrahs [*dharas*], and that other adventurers should be expelled'.[32] Many of the new additions might have been part-time soldiers and part-time agriculturists from Malwa villages. That turn of the century Pindaris were not necessarily outsiders and had strong local roots in the Malwa countryside is suggested by the ease with which they dispersed—dissolved is a more appropriate word—after an expedition, or when attacked.[33]

We have already noted that Pindaris had been customarily employed by the Maratha armies in Malwa. After 1805 Pindari participation in Maratha campaigns assumes a new significance. Yashwant Rao Holkar and Daulat Rao Sindia forged a sweeping alliance with the Pindaris, which was manifested in the Pindaris acquiring a limited share in political power. This provided the Pindaris an opportunity to enlarge the scope of their activities. Apart from being assigned specific territories, the prominent Pindari leaders were for the first time honoured with titles which announced their new status, and served to acknowledge their political clout in post-1805 Malwa. Chitu was honoured with the pompous title of 'Nawab Kamal Khan Mustaqim Jang', Karim Khan had the title 'Nawab Sarfaraz-ud-daula' conferred on him by Sindia while Dost Muhammad, another prominent Pindari chief, was also given the title of nawab.[34] After all, the Malwa states had to find a basis for the loyalty of the Pindari chiefs so as to ensure that the troops mobilized by them did not remain mere mercenaries. Had the third Anglo-Maratha War not intervened, these gestures might have become devices for incorporating some of the Pindari military leaders within the structures of the Malwa states.

Given the very low social status of the Pindaris, they had so far been kept at a distance.[35] Now for the first time, under Daulat Rao, Pindari leaders were deemed worthy of some social intercourse. To quote Malcolm,

> Although they commanded large bodies of men they were never allowed to sit down in the presence of the prince; and when Jeswunt Row [Yashwant Rao Holkar] . . . met Dowlet Row Sindia he reproached that prince for the encouragement he had given the Pindarry chiefs, by his personal intercourse with them, and by the high titles and grants of land which he had bestowed on men unworthy of such distinction.[36]

Nevertheless, Daulat Rao was not willing to admit any Pindari claim to sharing political power in Malwa on a permanent basis. Hence his anxiety to keep Karim Khan under control without completely alienating him. The company's residents at the Sindia court remarked that Karim did not really suffer rigorous confinement.[37] What is more, even in captivity Karim maintained close links with sections of Sindia's army particularly with the powerful faction led by Jaswant Rao Bhau.[38] Jaswant Rao was to provide shelter to Karim Khan and Chitu in the closing phase of the third Anglo-Maratha War.

The nexus between Jaswant Rao Bhau and the Pindari leaders is illustrative of the extent to which the military leaders of the Malwa states were dependent upon the Pindari soldiers. For instance in 1814 he was reported to have taken 'an immense body of Pindarries under the notorious Cheetoo' into service at the rate of Rs. 15 per month for each horseman.[39] Jaswant Rao played a major role in cementing ties between the Pindaris and the Sindia army during the inter-war period. Jaswant Rao was a leading commander during the first quarter of the nineteenth century. His father, Jiva Dada, had been one of Sindia's chief military officers in the late eighteenth century. Jaswant Rao held the parganas of Jawad, Jiran and Nimach (district Mandsaur, Madhya Pradesh) on the Malwa-Mewar border. These parganas had been taken by Sindia from Udaipur. Jaswant Rao had consolidated his position in this frontier region and used it as a base for encroaching upon the territories of states in south-eastern Rajasthan, especially Mewar. The fortified town of Jawad was the capital of Jaswant Rao's fiefdom.[40] He constantly needed military support to stabilize Sindia rule in the area. The ubiquitous Zalim Singh, de facto ruler of Kota for over four decades, was also active in this area and his intervention was often vital for Maratha expansion in Rajasthan. Zalim Singh was, among other things, an efficient military entrepreneur. He himself had an intimate connection with the Pindari chiefs. Zalim Singh assisted in negotiating the release of Karim Khan from Sindia's captivity. He facilitated the process by agreeing to stand surety for the huge sum (reportedly Rs. 6 to 10 lakhs) demanded by Sindia as the price of Karim Khan's liberty.[41]

During the third Anglo-Maratha War Jaswant Rao, ignoring the explicit instructions of Daulat Rao Sindia, actively aided Chitu and Karim Khan: 'In contempt of the orders publicly delivered to him on the part of Scindia, he not only furnished provisions, and every facilitation of progress, to the body of Pindaris retreating under Cheetoo, but he gave to that chief intelligence of the approach of the British troops, and saved him from being cut off.'[42] Seeing that Jaswant

Rao 'could not be prevailed upon to move a man against him', British forces attacked Jawad and captured the town, forcing the sardar to flee the place.[43] When the troops entered Jawad they discovered, much to their surprise that Karim Khan also had been sheltered by Bhau and had left shortly before the town fell to the British.[44]

Subsequently Jaswant Rao surrendered to the British (14 February 1818) and on the following morning Karim Khan also gave himself up.[45] Sindia punished Jaswant Rao's defiance by temporarily disowning him. Jaswant Rao's possessions, which had been occupied by the British, were restored to Sindia after the war. However Sindia refused to reappoint Jaswant Rao. The Bhau continued to petition Daulat Rao and it was only in 1821 that he was restored to favour. Even then, despite his appeals, he was not allowed to resume his former grant. Nimach, strategically located on the Madhya Pradesh-Rajasthan border, was made headquarters of the Rajputana-Malwa charge by the British in 1822 and placed under David Ochterlony. Ochterlony was Resident there till 1825. A cantonment for British troops was located on the outskirts of Nimach.[46]

Though Chitu seems to have been much more cautious than Karim Khan in that he did not overstep the bounds of his social origin in his relations with Sindia, yet between the two it was Chitu who symbolized much more forcefully the anti-British sentiment of the Maratha-Pindari coalition. This is not the place to go into details of the complicated end-game of the 1818 war. What one would like to recall is the almost simultaneous movement of Peshwa Baji Rao II in the direction of Asirgarh and the escape from British captivity and subsequent flight towards Asirgarh of Appa Sahib, the deposed ruler of Nagpur. Coinciding with this was the retreat of Chitu in the direction of Asirgarh. The Ahmadnagar Collector, H. Pottinger, reported in May 1818 that Baji Rao's force 'is said to consist of 7,000 men' and that Chitu and the Bhil Kanhya Bhil were with him.[47] In his final struggle Chitu aligned himself with Baji Rao, Appa Sahib and Kanhya Bhil.

Unfortunately the potential of this formidable Peshwa-Bhonsle-Pindari-Bhil convergence was never realized.[48] Chitu was forced to flee to the Vindhyan forests as British troops closed in, Malcolm was denied the satisfaction of capturing Chitu alive. Chitu was killed by a tiger in the dense Sita Ban jungle in Limanpur, not far from Satwas, where he had escaped. Malcolm sought the support of the local

Thakurs to apprehend Chitu. Bhim Singh of Bagli and Nana Zamindar of Kantaphor had offered to help in this task. Chitu was tracked 'like a hunted animal, through the jungles, by the prints of his horse's hoofs'.[49] Nana found the head of Chitu, 'the rest of the body having been devoured by a tiger', and presented it before Malcolm.[50]

Lying on either side of the Vindhya range, Bagli and Kantaphor (both in district Dewas) were part of the Sindia state. The road from Kantaphor runs in a south-easterly direction to Satwas and Nemawar. This entire area is not easily accessible. Malcolm described the 'fastnesses between Baglee and Mudleysir [Mandaleshwar]' as being 'chiefly inhabited by Gonds'.[51] Bhim Singh himself alluded to the fact that Limanpur could not be cultivated 'because it is a wild woody tract, the abode of thieves and robbers who have here and there inhabited a few hamlets'.[52] When a dispute arose in the 1840s over the sharing of Bagli revenues, reference was made to the service which Bhim Singh had rendered in cornering Chitu: 'Sir John Malcolm never forgot these services and always treated Bheem Singh with marked attention giving him presents of horses, swords, and guns'.[53]

The suppression of the Pindaris put an end to a major large-scale attempt on the part of the war-bands of central India, with their nucleus composed of warriors from the tribal belt whose strength was augmented by discharged troopers and impoverished Malwa peasants, to move out of the narrow confines of their domains and intervene in the high politics of the plateau. After 1818 the vast majority of those who had been members of these bands were still available for service in private armed contingents of petty-chiefs, landlords and powerful bankers. But the scale on which such contingents operated was too small to absorb all the former Pindaris. Thousands of these unemployed men with some military training became the mainstay of central Indian social banditry. In 1822 Ochterlony warned that 'the terror struck by the suppression of the Pindarry system is beginning to subside, and in some Quarters has subsided, and though the present Banditti have not yet assumed an open front, their aggressions are everyday becoming more formidable and more regularly organized'.[54]

The end of the Pindari model of tribal assimilation resulted in continued dislocation in the area. It is not surprising that western Madhya Pradesh retained a formidable tradition of popular resistance to colonial rule which, one would like to suggest, played a significant role in preparing the ground for the great revolt of 1857. This tradition earnestly sought to reverse the outcome of 1818. Tribal and forest communities, caught in the vortex of sudden changes caused by colonial intervention, remained an important component of this resistance.

NOTES

Abbreviations: CIA—Central India Agency Records; FD—Foreign Department; FD Misc.—Foreign Miscellaneous; *IESHR*—*The Indian Economic and Social History Review*; NAI—National Archives of India, New Delhi.

1. Canning quoted in C.H. Philips, *The East India Company 1784-1834*, Indian edition, Bombay, 1961, p. 215.
2. Guha, *Environment and Ethnicity*, p. 121
3. Cf. Surajit Sinha, 'State Formation and the Rajput Myth in Central India', reprinted in H. Kulke (ed.), *The State in India 1000-1700,* Delhi, 1995, pp. 304ff.
4. Cf. Guha, *Environment and Ethnicity,* pp. 110-16.
5. Dirk H.A. Kolff, *Naukar, Rajput and Sepoy: The Ethnohistory of the Military Labour Market in Hindustan, 1450-1850,* Cambridge, 1990, pp. 57-8.
6. Sumit Guha, 'Forest polities and agrarian empires: The Khandesh Bhils, c. 1700-1850', *IESHR*, Vol. XXXIII, No. 2 (1996), p. 136.
7. See R. Jenkins, Resident, Nagpur, 'Memorandum relative to the Pindarries', 1812, *Papers Respecting the Pindarry and Mahratta Wars*, NAI, FD Misc., 124/A, p. 26.
8. John Malcolm, *A Memoir of Central India, including Malwa and Adjoininq Provinces*, 2 vols, London, 1823, Vol. I, Chap. X.
9. James Tod, 'Origin, Progress and Present State [of the) Pindarras', Gwalior, October 1811, NAI, FD Misc., 124; Jenkins, 'Memorandum'; G. Sydenham, Agent, Berar, 'Memorandum respecting the Pindarries towards the end of the year 1814', 15 October 1814, NAI, FD Misc., 124. Cf. Edward Thompson and G.T. Garratt, *Rise and Fulfilment of British Rule in India*, London, 1934 (Allahabad, 1976 rpt.), Chap. IX; P.E. Roberts, *History of British India,* 3rd edn., Oxford, 1952, pp. 282-4.
10. The supposed reference to the Pindaris in the *Tarikh-i-Firishta,* is a misreading. Sydenham, using Capt. Jonathan Scott's translation (Firishta's *History of the Deccan*, London, 1794, Vol. II, pp. 121-2), stated that the Pindaris 'are first mentioned in Ferishta, . . . in his memoirs of Aurangzeb's operations in the Deckan, where he relates their having plundered the districts about Beejapore in conjunction with some rebellious Zemindars in the year 1689'. Sydenham, 'Memorandum', p. 37. Sydenham compounded the error by using the wrong era and thereby making Firishta, who resided at Bijapur from 1591 till his death in 1623, a contemporary of Aurangzeb. The mistake was repeated by Malcolm (*Memoir*, Vol. I, p. 426 and n.) and was carelessly reproduced in later writings. The notion that the Pindaris were

a product of the seventeenth century persisted in works such as those of G.S. Sardesai, which otherwise revealed a thorough grasp of Maratha history (G.S. Sardesai, *New History of the Marathas*, Vol. III, Bombay, 1948, p. 477).

It should be underlined that the Pindaris were essentially a phenomenon of the last quarter of the eighteenth century and more than a century separated the formation of this identity from Aurangzeb's Dakhan campaigns. Significantly Bhimsen's important memoir of these campaigns, *Tarikh-i-Dilkasha*, written almost a hundred years after the *Tarikh-i-Firishta*, does not mention the Pindaris (*Tarikh-i-Dilkasha*, English translation edited by V.G. Khobrekar, Bombay, 1972). Bhimsen was no stranger to the region which later became the main Pindari zone, since both he and his father had served in Malwa and Dakhan for a long time. Moreover, as Stewart Gordon observes, the Maratha identity and military ethos itself was still evolving during the seventeenth century (Stewart Gordon, 'Zones of Military Entrepreneurship in India, 1500-1700', in Stewart Gordon, *Marathas, Marauders, and State Formation in Eighteenth-Century India*, Delhi, 1994, pp. 193-4).

11. Malcolm, *Memoir*, Vol. I, p. 433 n. According to Malcolm, Karim Khan, one of the foremost Pindari leaders, stated that 'he had never heard any other reason given for this name' except that it was derived from 'an intoxicated [*sic*] drink termed Pinda'. It is significant that Karim Khan showed little interest in advancing a plausible theory about the origin of the Pindaris. He appears to have casually mentioned this etymology during the course of his interview with Malcolm shortly after his surrender, when he must have had more pressing problems on his mind.
12. Sinha, *Pindaris*, p. 2; Interview, Shri N.C. Zamindar, Indore, 1991 (Zamindar was scion of the *mandloi* family of Indore). Sardesai notes that 'the etymology of the Marathi word Pendha or Pendhar is doubtful. It means a body of stragglers . . .'. Sardesai, *Marathas*, Vol. III, p. 477, n. 8.
13. R.V. Russell and Hira Lal, *The Tribes and Castes of the Central Provinces of India*, Vol. IV, rpt., Delhi, 1975, p. 388. The Pindaris never became a fashionable subject for anthropological investigation since they were not a typical ethnic group. Further, they had no distinctive or exotic practices and beliefs which an anthropologist might have found exciting.
14. C.A. Elliott quoted in Charles Grant (ed.), *Gazetteer of Central Provinces of India*, 2nd edn., 1870 (Delhi, 1984 rpt.), p. xcv, n.
15. W. Irvine in an article entitled 'Etymology of the Word Pindari' (*Indian Antiquary*, Vol. XXIX, May 1900), argued that the term is derived from the word 'pandhar' which broadly referred to a region that included Hoshangabad, Nemawar and Handia on the Narmada. Quoted in Sinha, *Pindaris*, pp. 3-4. While tracing the early history of the Pindaris, Sydenham emphasized their connection with the Narmada region: 'in 1794 they are said to have established themselves on the Banks of the Nerbudda, and from that period to have increased progressively in consequence and power'. Sydenham, 'Memorandum', p. 38.
16. Jenkins, 'Memorandum', p. 30.
17. Sydenham, 'Memorandum', p. 46. Malcolm in his *Memoir* wrongly mentions Nimar instead of Nemawar as one of Chitu's possessions. This was apparently just a slip on the part of the author since he was too well acquainted with the region to make an error of this kind. See Malcolm, *Memoir*, Vol. I, pp. 444-5.

18. Jenkins, 'Memorandum', p. 30; *Central Provinces Gazetteer*, p. 28; P.N. Shrivastav, *Madhya Pradesh Gazetteers: Narsimhapur*, Bhopal, 1971, pp. 2, 54 nn. 1, 2.
19. *Narsimhapur Gazetteer*, p. 2.
20. Sumit Guha notes that in the first quarter of the nineteenth century 'the Gond territory extended as far west as Mandaleshwar on the Narmada river [district West Nimar], and eastward into Western Orissa and Chotanagpur'. Sumit Guha, *Environment and Ethnicity in India, 1200-1991*, Cambridge, 1999, p. 122.
21. Ibid., pp. 126-7.
22. Cf. James Tod, 'Journal', August-September 1813, NAI, FD Misc., 124, p. 30; Sydenham, 'Memorandum', p. 41.
23. 'Report of Commission . . . to enquire into Pindary invasion of Guntur, Masulipatam, Cuddappa', 31 January 1817, FD, Misc., 124/A, p. 56. This report, from which later historical accounts have quoted extensively, needs to be treated with some caution since it was very unusual for the Pindaris to operate so far down south. Moreover this was an isolated case which makes one wonder whether it was a Pindari raid at all. Nevertheless what is pertinent is that contemporary officials did not have a homogeneous group in mind while describing the Pindaris.
24. Tod, 'Pindarras', p. 6.
25. Ibid.; Sydenham, 'Memorandum', pp. 49-51.
26. Cf. Stewart Gordon, 'Legitimacy and Loyalty in Some Successor States', in Gordon, *Marathas, Marauders,* pp. 65ff.
27. Berasia (district Bhopal) was the birthplace of Karim Khan. Malcolm, *Memoir*, Vol. I, p. 449.
28. Ibid.; Capt. William Henley, Agent, eastern Malwa, to J. Stewart, Resident, Gwalior, 19 February 1819, NAI, FD, CIA, File No. 45; Sydenham, 'Memorandum', p. 51. For more details on the complicated tribute-sharing formula for north-eastern Malwa, involving Sindia, Holkar, Rajgarh, Narsinghgarh, Zalim Singh and the Pindaris, see Amar Farooqui, 'Towards Dussehra 1831: The Revolt of Lallaji Patel', *IESHR*, Vol. XXXV, No. 2 (1998), pp. 165-6.
29. Herbert Compton, *A Particular Account of the European Military Adventurers of Hindustan from 1784 to 1803*, London, 1892, pp. 31, 46-8, 229.
30. Jenkins, 'Memorandum', p. 26.
31. Jenkins talks of 'accessions which the result of the Mahratta wars must have brought them [the Pindaris], from the starving bodies of horse which the peace dispersed over Malwah'. Ibid.
32. Ibid.
33. Cf. Sydenham, 'Memorandum', p. 40.
34. Tod, 'Pindarras', p. 7; Sydenham, 'Memorandum', pp. 45, 55; Malcolm, *Memoir*, p. 441 n. 'Mustaqim' means faithful, firm.
35. Chitu, for instance, had originally been a slave. He was probably a Jat or Meo. Tod, 'Pindarras', p. 17; Sydenham, 'Memorandum', p. 44. The Pindari leaders never really managed to settle down and so were unable to manufacture impressive genealogies for themselves. Their humble origins are reflected in the commonplace names of a number of Pindari warriors: Chitu, Hiru, Rajjan, Gangola, Jangli, Kallu, Shukku, Nathu, Tukku, Buddan, etc.
36. Malcolm, *Memoir*, Vol. I, pp. 436-7.
37. Tod, 'Pindarras', p. 17.

38. Cf. Lord Hastings, Governor-General, 'Narrative of Military and Political Operations in Hindoostan and the Deccan', 1 March 1820, NAI, FD Misc., 124/A, p. 400.
39. Letter of Tod, dated 15 September 1814, NAI, FD Misc., 124, p. 88; also Cf. Sydenham, 'Memorandum', p. 55.
40. Letter to Lalaji Ballal, Kamavisdar, Jawad-Jiran, etc., 23 June 1802, Sri Nat Nagar Shodh Samsthan, Sitamau, Gulgule daftar, Vol. VII, No. 227; Madhya Pradesh State Archives, Bhopal, Sitole daftar (1805 onwards); A. Stirling, Secretary, Govt. of India, to Gwalior Resident, 23 May 1820(?). NAI, FD, CIA, File No. 106; NAI, Survey of India Memoirs, Route Book No. 36/11; Malcolm, Memoir, Vol. II, pp. 497-8; W. Hamilton, *East India Gazetteer*, Vol. II, London, 1828, p. 34; *Gwalior Gazetteer*, pp. 245-6, 265, 275 and 293.
41. Bengal Govt. to Secret Committee, 16 August 1811, NAI, FD Misc., 124/A, p. 1; Tod, 'Pindarras', p. 12; Sydenham, 'Memorandum', p. 51; Malcolm, *Memoir*, p. 455. It is not unlikely that Zalim Singh made a large profit out of this guarantee. He was an astute banker and was engaged in various commercial dealings, including the supply of opium. One suspects that some of the Pindari chiefs might have maintained accounts with Zalim Singh and that he was willing to stand surety for such a large sum precisely because he had access to Karim Khan's deposits. Given the fact that the Pindaris led a rather insecure existence and were constantly on the move, they had to rely on the services of bankers to look after their financial arrangements. In the absence of a regular treasury this was indispensable. We learn that Dhakun (Dakhan?) Seth was the main banker of Chitu's camp. Another banker, Mowjee Ram, was also attached to Chitu. These bankers and sahukars advanced money to the Pindaris for the purchase of horses and food and were repaid from the proceeds of tribute collections and raiding expeditions. Sydenham, 'Memorandum', p. 49.
42. Hastings to Court of Directors, 8 February 1818. NAI, FD Misc., 124/A, pp. 204-5.
43. Hastings, 'Narrative', 124/A, p. 400; J. Blacker, Quarter-Master-General, to Major-General W.G. Keir, 6 February 1818, NAI, FD Misc., 124/A, p. 332.
44. Hastings, 'Narrative', pp. 400-1.
45. Malcolm to F. Warden, Secretary, Bombay Govt., 17 February 1818, NAI, FD Misc., p. 238.
46. NAI, Survey of India Memoirs, 36/11; *Gwalior Gazetteer*, p. 275.
47. H. Pottinger, Collector, Ahmadnagar, to M. Elphinstone, 16 May 1818, FD Misc., 124/A, p. 285.
48. In his analysis of Bhil motives in being collaborating with this subversive coalition Guha states that the Bhils 'were not isolated remnant populations savagely defending themselves against inexorably encroaching civilization—rather they were determined to share in the goods that civilization could yield, and to secure this they were prepared to use their mutual solidarity, their knowledge of the hills and forests, of the paths and water-resources, their capacity for guerrilla war, as assets to be traded in the regional political arena. They could be allies of the Mughals, of Holkar and offer refuge to the last Peshwa himself'. Guha, *Environment and Ethnicity*, p. 110.
49. Malcolm, *Memoir*, Vol. I, p. 447.

50. Ibid., pp. 447-8; R.N. Hamilton, Resident, Gwalior, to H.M. Elliot, Secretary, Govt. of India, 22 July 1850, NAI, FD, CIA, File No. 129.
51. Malcolm, *Memoir*, Vol. I, p. 522 and n.
52. Thakur Salim Singh and Kunwar Bhim Singh of Bagli to Maj. W. Henley, 1877 (Samvat), tr., NAI, FD, CIA, File No. 129.
53. Hamilton to Elliot, 11 July 1850, NAI, FD, CIA, File No. 129.
54. David Ochterlony, Resident in Malwa and Rajputana, to G. Swinton, Secretary, Govt. of India, 30 April 1822, N.K. Sinha and A.K. Dasgupta, ed., *Selections from Ochterlony Papers (1818-1825),* Calcutta, 1964, p. 244.

Forest-Human Interface: The Forest Policy of the Native State of Cochin (1800-1947)

K.T. Thomas

The forests of the native state of Cochin were mainly situated in the eastern part of Talappilli, Trichur and Mukundapuram taluk and southern part of Chittur taluk. In the nineteenth century there were some forest disputes between Cochin and Travancore, and some other chieftains. Most of these disputes were settled in favour of the Cochin state. In 1915, the Valia Nambidy of Koillengode claimed Govindamala in Nelliampathies and a compromise was arrived at in 1922 by giving the *jenmam* title of the area to him under the Cochin government. In 1920, the Kavalappara Moopil Nayar claimed a portion of the Medugal forests and a settlement was made by assigning to him an area of 666.44 acres of *jenmam* and *verumpattam* (simple lease) tenure by excluding the area from government reserve forests.[1]

The most valuable trees of the Cochin forests—teak, ebony and rose wood—were government monopolies.[2] Teak had been royal monopoly from very early times, but ebony and rose wood were made such later.

II

The forest administration of the Cochin state was managed in a way to extract the resources as much as possible. The early practice was one of leasing the forests to contractors for specific periods for a lump sum.[3] The whole forest area was placed at the disposal of the lessees, with no limit being placed on the quantity of timber to be annually removed.[4] The government was not satisfied with this system for it did not bring in the expected revenue. Therefore, in 1813 Colonel Munro (the Resident) abolished this practice and made arrangement by appointing a Malamel Vicharippu or Superintendent of the Forests, with required staff. Their chief function was confined to the extraction of teak trees. Private individuals could remove all kinds of trees (except monopoly trees) and minor forest produce on payment of customs and transit duties. This arrangement resulted in a three-fold increase in the income from forests. The appointment of Lieutenant Lethbridge as the Chief Superintendent in 1818 led to large-scale extraction of teak from the forests to tide over the financial problem.

The government had advised the forest managers to extract and store enough timber when there was much demand.[5] They were asked to identify ebony trees which were best for shipbuilding. Large quantities of such trees were to be cut and stored to meet the growing demand.[6]

The Cochin government had a problem in bringing the timber from the northern forests to the depots. The timber had to pass through Edathuruthy, which was a British territory, where transit duties were collected. In 1808 large quantities of timber had been blocked there by the Company officials demanding an extra duty of Re. 1 per candy.[7] The problem was later settled. In 1815 an agreement was entered into for mutual relinquishment of inland duties on timber which were floated down the rivers common to the Cochin and British territories.[8]

In 1835 J.A. Kohlhoff was appointed Conservator of Forests. He divided the forests into seven districts, each of which was placed under a Vicharippukaran, assisted by other staff. While the monopoly trees continued to be extracted departmentally, the permit system was introduced for extracting all other trees and the contract system for the removal of firewood and minor forest produce.[9] This system continued till the end of the nineteenth century. Private individuals were allowed to extract trees (except monopoly trees) on the basis of the 'permit system' by paying the prescribed *kutikanam* or seigniorage fee.[10]

These administrative measures helped the government to extract the resources on a large-scale. The income from the Cochin forests rose from Rs. 25,978 in 1871 to Rs. 3,11,382 in 1900.[11] But the system of management was corrupt and inefficient. It was worked by low paid, uneducated and untrained subordinate staff. Even in 1940 the salary of a first grade forest guard was only Rs. 30 and that of a last grade was Rs. 18.[12] There were several instances of supervisors of forests engaged in corrupt practices. Some of them even sold government owned teaks to private individuals and made personal profits.[13] The concept of forest conservancy was alien to these low paid and uneducated staff. The work of timber extraction was carried out in a very indiscriminate and reckless manner by the officers as well as by the permit holders. This led to the depletion of the accessible forests.

In the 1890s when income began to show a declining trend, the government secured the services of a forest expert from the Madras government. The appointment of V. Alvarchetty led to the reorganization of the department. Unregulated fellings by permit holders under the *kutikanam* system was gradually put an end to. The fellings were confined to trees marked by the superior officers and the work was given on contract to the lowest bidders.[14] The income from the forests substantially improved during this period. However, the exploitation of the accessible forests continued without giving any rest to the over-exploited areas.

There was no written forest law till 1905. Regulation 111 of 1080 M.E. was enacted in 1905 on the lines of the Madras Act V of 1882. This regulation was amended thrice by the Regulation VII of 1918, IV of 1920 and I of 1925.[15] These rules regulated among other things the demarcation and settlement of forest areas, the protection of reserved and unreserved areas, the felling and removal of monopoly trees, the transport and floating of timber, and hunting, shooting and fishing, etc., in the reserved forests.[16]

For want of transport facilities, the forests known as Orukomban Working Circle were not exploited till 1906. To extract timber from the inaccessible area, a forest tramway scheme was prepared. Started in 1901, the tramway was completed in 1906 and was opened to traffic. For the protection and management of the steam tramway, the government passed a regulation.[17] The tramway was made a separate department which was finally merged with the Forest Department in 1938.[18] The tramway consisted of three sections with a total length of 50 miles. The main block of the forests worked by the tram line from 1907 to 1928, yielded a revenue of Rs. 69,91,043.[19] It rose to Rs. 1,36,88,879 by the end of 1945.[20] When the quantity of timber in the forests tapped by the tramway began to get reduced,[21] the government began to worry about the maintenance of the tramway which was very expensive. To keep the tramway going it was therefore decided to work down all the teak and rose wood toplings of the worked coupes and also extract and transport timber from as many new coupes as possible.[22] Thus, even for the sake of maintaining the tramway, the forests were subjected to over-exploitation.

In 1944-5, the total timber derived from the forests through the department and private agencies was 31,407 tons, firewood 65,011 tons, and bamboo 45,900.[23] The revenue receipts from forests in 1945-6 was Rs. 33,13,070 with a net surplus of Rs. 14,66,991.[24]

III

Besides the extraction of timber by government and individuals, another cause of deforestation was the encroachment of forest area by cultivators of fugitive crops. After the construction of irrigation works in Chittur taluk, almost the whole of the teak area there was assigned for cultivation between the years 1855 and 1875. The acting conservator reported that in the year 1871-2, 2,000 *parahs* of lands were cleared for cultivation but no yield could be expected for the next three years or more.[25] The government did not prevent the encroachment of forest areas by cultivators.

The government also assigned forest areas to the farmers for food crops cultivation. To produce more food crops to keep pace with the growth of population, the government began in 1920s, to make available more forest areas for cultivation. Thus in the first half of the 1930s about 15,000 acres of forest lands were assigned for food crops cultivation.[26] The breaking up of forest lands for cultivation and the burning of even valuable timber for manure, both of which were ruinous practices, were not interfered with.[27] Anticipating a food crisis, an area of 7,057 acres of reserved forests was brought under agricultural operations on temporary leases during 1945-6 as against the extent of 5,000 acres in the previous year.[28] The government considered it an achievement of bringing in extensive forest areas under food crops cultivation, especially rice.

The state also maintained village reserves bordering on reserved forests. In 1945-6 the total village reserve was 2,689 acres and 28 cents.[29] These areas had been set apart exclusively for catering to local agriculturists' requirements in the matter of grazing, supply of fodder, green leaf manure, firewood, small timber, fencing materials, etc. The government used to lease few hundred acres of reserved forests for the occupation of the depressed class families. These areas were excluded from the reserve forests and were treated as revenue *porambokes*. The total extent of the reserved forest leased under this category at the end of 1945-6 was 319 acres.[30]

The proliferation of the tile factories in Trichur area from the beginning of the twentieth century enhanced the demand for forest fuel. Availability of cheap firewood in the vicinity was a major factor for the rapid growth of this industry in that region. This also added to the devastation of forest wealth.

Cochin was not under direct colonial rule. Nevertheless it had accepted British paramountcy by the treaty of perpetual friendship. This affected the Cochin state in various ways. The annual tribute payable to the English put a heavy financial burden on the tiny state of Cochin. This resulted in an increase in taxes and indiscriminate exploitation of the forest wealth. Almost throughout the nineteenth century the Forest Department was headed by British officers like Lethbridge and Kohlhoff in collaboration with the British Residents. Their policy was to improve the efficiency of the forest administration to extract more resources. That policy was in tune with the colonial approach towards forests and conservancy.

In the second half of the nineteenth century Cochin forests, especially Nelliampathy Hills, attracted foreign capitalists. Between the years 1862 and 1879 about 9,420 acres were leased to various companies and private individuals. There were 17 coffee estates aggregating 8.251 acres in 1882-3.[31] The total amount of rent of these estates was Rs. 8,155 and 8 *annas*.[32] Later the number of estates rose to 19 of which only 1 was owned by a native; all the rest belonged to foreign capitalists.

Forest areas were also opened for rubber plantations in the beginning of the twentieth century. Rubber plantation was first started in Palapilly in 1905 which was soon followed by another in Vellanikara.[33] Encouraged by the success, further areas in Palapilly and Sholayar were thrown open for rubber plantation.[34] These were purchased and planted by foreigners. A native company was also allowed to take up 1,000 acres for rubber plantation. In 1936 there were 9 rubber estates paying an annual quit rent of Rs. 25,852.[35] Tea and cardamom cultivation were also encouraged at the cost of reserved forests. The extent of reserved forests sold or leased out to private parties for the purpose of raising hill produce, such as rubber, tea, cardamom and coffee, was 13,671 acres at the beginning of 1945-6. An areas of 547 acres were leased that year and thus the total extent of area sold or leased stood at 14,218 acres at the end of the year.[36] The European capital investment in the plantation sector led to the devastation of the forest wealth.

The demand for sleepers from the Cochin Railway and Madras Railway Company encouraged the government to cut down tress indiscriminately. Monetary gain was the prime motive for the government.[37] The government decided to exploit Machad forests for locomotive fuel. A contract was signed with the Madras Railway Company for a monthly supply of 100 tons, which was then raised to 550 tons[38] in 1899-1900; 6,949 tons of locomotive fuel were supplied to the above said company. The demand for rail sleepers also continued and 2,880 sleepers were supplied in 1938-9.[39]

The World Wars in which Britain was a major partner, further escalated the demand for forest produce. The war situation created favourable market conditions for both valuable timber and inferior soft wood species. The opportunity was fully availed of and the revenue receipts from the Forest Department almost doubled in 1943-4.[40] The cessation of World War II necessitated a switch-back from the very intensive programme of war supplies to something more normal, but market conditions had not settled down because of the need for post-war reconstruction. Thus both the war and post-war situations heavily taxed the forests of Cochin. In short, colonial strategies and policies contributed directly and indirectly to the depletion of the forests of Cochin.

IV

It was only in 1872 that the government made the first attempt at reforestation. Teak planting was commenced during the year on the Palapilly Hills, and only 10 acres were cleared and planted. Between 1872 and 1888 the total area of the new plantations rose to 835 acres. In 1882-3 the revenue receipts from the Cochin forests was Rs. 1,21,533 while the amount spent on reforestation was only Rs. 5,791.[41] Anticipating a shortage of teak which was expected to occur through the depletion of natural forests, teak plantation was started in Chettikulam in 1912. An experiment on sandalwood plantation was made which proved successful. The new teak plantations undertaken in the Northern and Southern Divisions between 1901 and 1928 covered only 2,041.44 acres at a cost of Rs. 32,860.[42] Reforestation was mainly confined to teak plantation which slowly rose to 10,552 acres by the end of 1945-6.[43] But government spending on teak plantation considerably declined to Rs. 84,578 in 1945-6 against Rs. 2,49,172 in the previous year.[44] While the Forest Department earned a revenue of Rs. 33,13,070 in 1945-6, it spent only a few thousand rupees on reforestation.[45] The only positive approach towards conservancy could be seen in fixing the minimum girth of the exploitable tree as 6 feet.

The reforestation attempt of the Cochin government was not influenced by any concept of conservancy or environmental consideration. Commercial consideration was the chief motive behind reforestation. Indiscriminate extraction of forest resources by government and private agencies and encroachment by cultivators depleted the reserve forests. While several thousands of acres of reserved forests were set apart for plantations and food crop cultivation, only few thousand acres were newly planted with teak trees. Even for planting teak existing natural forests were cleared and felled.[46] While the revenue from the forests increased manifolds, the government spent only a meagre amount on reforestation. It did not take adequate measures to prevent forest fire which destroyed many valuable trees. The only block specially protected from fire was the Palapilly teak plantation, an area of about 800 acres. Existing forests were over-exploited without giving any rest. In short the economic importance of forest conservancy and ecological considerations did not influence the forest policy of the Cochin state.

V

The Cochin government viewed the forests as a continuous source of income to the state. The establishment of a separate department and the subsequent administrative measures to improve the efficiency of the department were aimed at maximum extraction of the forest resources. The large-scale exploitation of the forests for capitalist plantation, food crop cultivation and the low priority and negligible investment on reforestation betray the notion that the native states were ecologically friendly. Even the reforestation, done after deforestation, was with a view to ensuring the supply of valuable timber in future.

Colonialism had its role in the deforestation of Cochin. Acceptance of the British paramountcy put a heavy financial burden on the small state of Cochin. To mobilize the annual tribute to the British, the Cochin government was forced to enhance taxes and extract the forest resources indiscriminately. The appointment of British experts at the top of the forest department; the colonial capital investment in plantations, formed after clearing the natural forests; the contract with the Madras Railway Company for the supply of sleepers and locomotive fuel; and above all the favourable market conditions created by the World War situations, all led to the devastation of the forest wealth.

This paper puts forward the following conclusion:

(1) It refutes the concept that during the colonial period, the native rulers had an altogether different approach with regard to forest policy and conservancy.
(2) The colonial policy had its effect on the depletion of forests in Cochin.
(3) Even the limited attempts of reforestation and conservancy was not influenced by ecological considerations but by economic or commercial motives.

NOTES

1. T.K. Krishna Menon (ed.), *The Progress of Cochin* (Ernakulam, 1932), p. 141.
2. Regional Archives, Ernakulam (hereafter RAE), MSS, Series Files II, No. 1016, 1815.
3. C. Achyuta Menon, *The Cochin State Manual*, p. 253.
4. Ibid.
5. RAE, MSS, A Copy of Regulation to the Managers of Forests, Series File No. 1082, 1814.
6. RAE, MSS, Series File II, No. 1008-14.
7. RAE, MSS, Series File I, No. 76, 1808.
8. RAE, MSS, Series File III, No. 2, 1825.
9. C. Achyuta Menon, op. cit., p. 254.
10. Ibid.
11. Government of Cochin, *Report on the Administration of Cochin State* (hereafter *RAC*), *1871-2* (Ernakulam, 1873).

12. Government of Cochin, *Record of Administration*, Vol. I (Ernakulam, 1941), p. 67.
13. RAE, MSS, Forest Book, 1883.
14. C. Achyuta Menon, op. cit., p. 25.
15. Ibid., p. 260.
16. Ibid.
17. Government of Cochin, *Regulation V of 1082, Cochin Forest Tramway Regulation*, (Ernakulam, 26 June 1907).
18. *Record of Administration*, p. 62.
19. Krishna Menon (ed.), op. cit., p. 62.
20. Government of Cochin, *Report of Administration—Forest Dept., 1945-6* (Ernakulam, 1947), p. 51.
21. See the Budget Estimate of Cochin State 108ME (1912), p. xi and 109ME (1917), p. viii.
22. Krishna Menon (ed.), op. cit., p. 146.
23. Government of Cochin, *Report of Administration—Forest Dept., 1945-6*, p. 7.
24. Ibid., p. 19.
25. *RAC, 1871-2* (Ernakulam, 1873), p. 25.
26. Krishna Menon (ed.), op. cit., p. 146.
27. *RAC, 1898-9* (Ernakulam, 1900), p. 38.
28. *Administration Report—Forest Dept.,* 1945-6, p. 3.
29. Ibid., p. 1.
30. Ibid., p. 2.
31. *RAC, 1882-3*, p. 16.
32. Ibid.
33. Krishna Menon (ed.), op. cit., p. 147.
34. Ibid.
35. Ibid.
36. *Administration Report—Forest Dept., 1945-6*, p. 2.
37. RAE, MSS, No. 880, Forest Book, 1883.
38. *RAC, 1898-9* (Ernakulam, 1900), p. 38.
39. See Government of Cochin, *Administration Report—Forest Dept., 1938-9*, p. 20.
40. *Administration Report—Forest Dept., 1943-4*, 1945, p. 24.
41. *RAC, 1882-3*, p. 47.
42. Kirshna Menon (ed.), op. cit., p. 150.
43. *Administration Report—Forest Dept., 1945-6*, p. 10.
44. Ibid.
45. Ibid., p. 19.
46. RAE, MSS, No. 998, 1889.

Tribal Resistance in Chotanagpur: A Case Study of the Dubia Gossain Movement (1870-1880)

P.K. Shukla

In spite of several scholarly works[1] produced by the ethnographers, anthropologists and historians on the tribal movements from the Kol insurrection in Chotanagpur to Tana Bhagat movement (1914-21), there is still a crucial gap in the study. There is hardly any serious work which exclusively narrates these movements in the background of colonial expansion in the Santal tract and links them up with the expropriation of the surplus generated through the imposition of land rent, its exorbitant enhancement and control of the forest wealth by the colonial authorities, and at times their collaborators such as the land-grabbers, the mahajans and the moneylenders. Such exploitative measures resulted from historical circumstances as well as the shape and form of the British revenue settlements.[2] The census operation, enumeration of the houses, preparation of the records of land rights, fixation of land tax and categorization of the land in various categories, all such measures and enactments of various laws in the 1860s and 1870s aimed at imposition of land tax and fixing enhanced rent so as to accumulate surpluses for furtherance of the interest of the British Empire in India and the East. Most of the Santal tribes were landless agricultural labourers as they were evicted from their traditional landholdings. Those who cultivated the landowner's land with no restrictions on rent were tenants at will.[3] The changing forms of appropriation by the creditor were noticed during this phase of intensive colonialism. The problem was most acute in the tribal areas. For instance, the Santal of the Damini-i Koh, under the police on the one hand and the civil courts on the other, as manipulated by mahajan adventures from Bhojpur in the Shahabad district, who systematically exploited them, were reduced to a condition of serfdom in which they could not buy a bullock, an utensil, or even a grain of the produce of their own. Though the revised Civil Procedure Code of 1877 prohibited the attachment by creditors of the 'implements of husbandry', the peasants could still be pressurized to sell their articles in order to pay the creditors. It has been established that during 1883-4 in the Chotanagpur division 447 land transactions took place in which 374 ryots were affected. Again, during 1892-3, 1516 transactions/ mortgages of land took place in which 1289 ryots suffered. Land sale also increased remarkably in the Santal pargana. Although Regulation III of 1872 recognized tenant rights and also guaranteed security of tenure and fixity of rent, in effect, it was of no avail to restrict interest to 24 per cent. Within ten years of the settlement, 10,000 court sales and 40,000 private sales of ryoti holdings have been noted. The pace of colonization of the tribal regions can be judged from the incidence of immigration. In Chotanagpur alone, for instance, between 1871 and 1931, the immigrants rose from 96,000 to 3,07,000.[4] There were a few Santal peasants who paid revenue directly to the state or paid rent that was fixed, or could be increased only within limits fixed by law. This type of fixed regulation for collection of land revenue and payment of rent was never prevalent in the Chotanagpur areas. British rule and its accompanying commercialization strengthened the already present tenancies towards penetration of tribal areas by outsiders from the plains—moneylenders, traders and land-grabbers and the contractors. British legal conceptions of absolute private property thus eroded tradition of joint ownership such as the *khuntkatti* tenure in Chotanagpur. A new but increasingly important factor from the 1870s and 1880s was the tightening of control by the colonial state over forest zones for revenue purposes.[5]

It was because of these factors that the tribals of Chotanagpur launched a series of protests against the British rule and the *dikus* (outsiders), though the forms of protests were 'a curious mix of agrarian, religious and political issues'. They were followed by socio-religious of revitalization movements.[6]

In order to understand the real nature of the Dubia Gossain movement (1880) it is necessary to discuss briefly the Kherwar movement (1874) because in the heart of their hearts the Santals or

Kherwars (the ancient tribal name of the Santals) cherished an agrarian dream which was strongly connected with the Santal raj. To them, Kherwarism was inseparably associated with the golden era of their history when they lived in Champa in absolute independence and had no rent or tribute to pay but only a small offering to their leaders by virtue of their office.[7] Since the so-called golden days in Champa, the Santals felt that they had no opportunity to act en bloc, but the settlement in the Damin area and adjoining districts made it possible. During all these years the dream of the Santal raj had been kept alive, and when leaders arose they were ready to act. Though their knowledge of the 'mighty power' of their rulers was very limited, and so was the textual knowledge of the Hindu scriptures, the passionate appeal to the past as well as an ideal to shape the present on the basis of the borrowing from the dominant Hindu culture to improve their economic condition was very much in their minds. When any reform initiated by the government failed to satisfy the Santals, they raised the standard of revolt in 1860 and 1871.

The Santals could not be fed for long on mere false promises. The movement of 1871 in the Santal pargana had hardly died down when in the autumn of 1874 there was yet another serious outburst of the Kherwar movement under the leadership of Bhagirath Manjhi of the village Tardiha in the Godda subdivision. He had also taken part in the rebellion of 1855. He was imprisoned in 1868 for seditious conduct in trying to disturb people's mind by threatening an outbreak. Stephen Fuchs has rightly characterized the reasons behind the Santal's agitation. He observed that the outcome of the excitement among the Santals was partly due to the settlement operation. Disaffection manifested itself in resistance to the payment of rent and attempts to form a kind of political organization. Bhagirath Manjhi and Gyan Parganait were the moving spirit of the rebellion. Stephen Fuchs says:

> The tradition of former purer belief ordinarily does not trouble the Santals, but the dormant memory of God is awakened when anything extraordinary happens, a famine or epidemic, for instance. Then the Santals feel guilty of abandoning their supreme God and worshipping the spirit, and they vow to reform themselves. This also explains the spasmodic character of the movement. In times of comparative plenty and prosperity very little is heard of it, during times of distress the movement revives.[8]

The Bhagirath Manjhi movement of 1874 occurred in the year of great scarcity and famine. The Santals were in great distress. He appeared with his Kherwar idea by which he wanted to attract the attention of the hungry masses among the Santals. But in some places people, not Kherwars, were found mixed up with actual 'obstructionists'.[9] It may be noted that other non-Santal oppressed people such as the Kurmis, the Koeris, the Kahars and the Momins participated in the movement, although their gods and goddesses were different. Bhagirath was the leader whose teachings assumed more and more political slant when he began to proclaim that the land which the Santals cultivated belonged to them by right and that no government could demand taxes from them. In his meetings he asserted that the land which they and their ancestors reclaimed and made cultivable belonged to them. To suppress the movement the government took strong action to check it. Ultimately, Bhagirath was arrested and sentenced to two years imprisonment. The Deputy Commissioner, J. Boxwell demolished the shrine set up at Taldiha (Godda), carried away the image and took the Santal Pandit, Bhagirath's brother Ram Gyan Parganait and Hindu Matadin with him to Dumka.[10] An extra detachment of troops from 4th Native Infantry from Bhagalpur was stationed at Dumka and new police posts were established. Ultimately, the troops were withdrawn in 1877 when the agitation was thought to have subsided. Bhagirath was released in 1877. Though he settled down to become a quiet and peaceful religious reformer, he continued his work in a secret way. It was discovered later that the movement did not die, but that there did exist a distinct organization of Kherwars which was silently spreading and strengthening it. Bhagirath died in 1879 but a new movement against the census operation of 1881 was launched, and it was more violent than any other movement since the Santal Revolt of 1855.[11]

Dubia Gossain Movement (1880)

This movement was an outcry against the census operation which was raised by the peasants of Jamtara under the influence of Dubia Gossain. It was initiated by him and it was eagerly caught up elsewhere. The Kherwar leaders seized the opportunity quickly and headed it, which advanced Kherwarism with the strength of a revival. Even the commissioner was of the view that as a result of the census operation, the Kherwar organization had made a great gain both in respect of adherents and the strength of their principles.[12] The revival of the movement was considered by the government as dangerous, contemptible and not innocent. Its only exponents in Pakur were Bhagrai and scribe Lakhan. Their manifestations were confined to complaints and petitions containing long expostulations on local matters such as the rates of rent, or the system of *chaukdari* payment. It may be noted that Damin-i Koh was re-settled with an increase of 80 per cent in its revenue caused by the assessment of new cultivation. It was reported that some five men, all under the influence of *ganja* appeared before the subdivisional magistrate of Dumka with some fragments of papers, scrawled with unintelligible characters, which they claimed had fallen from heaven. They said that they had been directed by the deity to take them to the magistrate. The Deputy Commissioner of the Santal pargana enquired into the matter and found out these men were aggrieved at the disappearance of the village grazing land under the operation of the settlement.[13] Both the divisional officers of Dumka and Deoghar enquired into the matter and found out no other reason than the anger of the people caused by settlement operation. With due warning these men were discharged.

The new disturbances had its seat outside the Santal pargana, at Jagesar in the Hazaribagh district of Chotanagpur. The new messiah Dubia Gossain had a thorough knowledge of the Santal character and their Kherwar heroes. The saint had travelled through several districts in which Santals were located and by austerities which he practised succeeded in attracting the notice of the people. He had three Santal agents, one of whom had been employed in the police and could write a little. Those men spread all sorts of rumours in the district. It was therefore natural that the wonderful accounts of Babaji's supernatural powers were widely disseminated. These ran their course like wildfire and caused the utmost consternation among the ignorant. They were followed up and intensified by a volley of letters which were passed on from village to village. Verbal messages were also passed on through whispering, ordering slaughtering of all pigs and fowls. Giving up drinking, forsaking of all such deities and conforming to the usages and customs of the Hindus were preached. To enforce obedience it was predicted that calamities such as deluge of fire, rain, etc., would overtake all those who neglected the rituals. A promise was also given that if they were in readiness by the Dussehra festival, the Santal raj would then be made over to the followers of Babaji, and that there would be no more rack-renting of oppression. Threats against the mahajans were not wanting and the new era held out little prospect of wealth and happiness for them.[14]

In Jamtara division lay an important Ghatwal taluk called Narainpur. It was under the Court of Wards which was mismanaged. It was the site of a large weekly market which the Santals attended and discussed the approaching census in a spirit which excited alarm among their own Santal neighbours. The missionaries refused to undertake the census operation as they were told by the Santals that they would kill any enumerator who came to their villages. Only Bengali agents of the taluqdars were asked to do so. An officer of Jamtara once attempted to retain temporarily the ring leaders of the Santals, the crowd threateningly called for their release. Ultimately, the Deputy Commissioner of Jamtara met the ring leaders alone at Karmatar. They promised that they would bring their headmen. They came with a tumultuous and threatening mob who raised rebellious cries. The Deputy Commissioner was secretly apprised by the *perguanits* of the character of opposition, the language used by the Santals. But the zamindars of Jamtara supplied armed men and the ring leaders were arrested. The crowd sullenly broke up but the Deputy Commissioner's house was set on fire the same night. Fifteen armed police arrived from Bhagalpur and the enumeration was begun and pushed.[15] This violent action of the Santals was influenced by Gossain Baba. House numbering was another step to be undertaken by the colonial authorities when the census was on. The Kherwar *mustajir* in Damin-i Koh frankly told the Assistant Commissioner of Paku, Mr Campbell that he would not allow houses to be numbered. When he persisted in his defiance, he was arrested and sentenced to three months' imprisonment.

It was reported that opposition to the census operation was found in Khairabari, Gumru Ghatwal taluk and along the border of Dumka. Another Baba named Karu Santal also emerged at Jhemakur in the Jhertal taluk of Handwa. The new Baba was also called Tribhuban Raut. The Santals of a small taluk resisted their tahsildar, who was the enumerator and was compelled to return to the written schedules which he had already prepared. Two or three cases of opposition to the operation by the Kherwars were reported from Rajmahal and a crowd of some 500 Santals had gathered there. In the western taluk of Handwah Taljhari the property of a non-resident and alien zamindar was also in the census charge of the Deputy Inspector of a school. A strong opposition was noticed there and 50 Santals assembled there. A meeting with the Deputy Commissioner and the ring leader, Raghai was arranged. Raghai took note of the party of 25 armed police who attended on him. He went away with his followers and in the afternoon returned with a strong crowd of 400 Santals, each armed with a short staff. The Deputy Commissioner found their spokesmen all in opposition and after forming them into panchayat attempted to enumerate Raghai's household. The latter refused to give the particulars and was arrested. The Santals then announced their determination of accompanying Raghai wherever he went. Ultimately, the Deputy Commissioner ordered for the 11th Madras Native Infantry which had halted 6 miles further on the road to Deoghar. The crowd followed and eventually dispersed but they reassembled for the next four days, when finally Raghai announced that he would get the whole Taljhari taluk enumerated.[16] Although Handwah was the centre of opposition, the disturbances against census were noticed in Nanhati, Dumka, Godda, Katikund, Deoghar and Damin-i Koh. It was owing to the presence of troops of cavalry that the officials of the census were able to move from one taluk to another.

It was Kharwar idea which attracted his followers to Dubia Gossain and gave him influence. According to Barlow, the officiating Commissioner of Bhagalpur and Santal pargana, Dubia Gossain had a large influence among the 16,000 Santals of the entire Deoghar and Jamtara jurisdiction and the border of Dumka subdivision. It was reported[17] that during the census operation, the Kherwar gained a host of new followers. In order to suppress the influence of Dubia Gossain the government took punitive action and Dubia Gossain was taken prisoner and was deported to Lucknow, far away from the Santals, as a political prisoner. Other ring leaders were also arrested during and after the census operation and sentenced to either imprisonment or deportation. The Manjhis, who had had any connection with the Kherwars, were removed from their official position, and the whole question of Santal administration was taken up for reconsideration. In order to break down the movements of the Santals, the Deputy Commissioner of Santal pargana proposed to admit outsiders into Damin area and to introduce the regular police into the whole of Santal pargana.

From a close scrutiny of the sources available on the Dubia Gossain movement, it appears that it was mainly directed against the colonization of the area. It was an agrarian issue and not a religious one, because the demands of the Santals included the restoration of their land, opposition to enhancing rent and encroachment to their grazing land and forest wealth, although they made use of the religious myths, legends and superstition for uniting the people to launch a struggle for their economic betterment. It was admitted even by the colonial authorities that 'they were at first allowed to deal with the forest as they liked, not only in the Damin-i Koh, but in the rest of district into which they were introduced early in the present century'.[18]

In order to find out the nature of the movement a series of twelve questions was sent to all the colonial authorities and the zamindars of the areas. Almost all of them confirmed that Dubia Gossain's movement was widespread. Maharaja Gopal Chandra Singh of Maheshpur reported that the opposition to census did spread rapidly. Babu Taresh Nath Pande of Pakur reported that the Santals objected to the counting of houses taking place at night. Apart from actual opposition there were rumours everywhere near Hazaribagh. The Santals did not want the names of sister-in-law and brother-in-law to be registered together. They apprehended that the men were to be sent as coolies in some war; some said that new poll tax was to be levied and that all able bodied men were to be sent to Chchar to work in the tea plantation. Some Santals said the government was going to recruit them in the army to be sent to Kabul in the Afghan war. Babu Ram Kanai Karforma admitted that low classes of Hindus, the Bhuiyas, and Goalas, Kols, the Chamars, the Jolahas, the Doms, the Koeris and the Kurmis—all of them objected to the census operation. Several excuses were made by the Santals but in the heart of their hearts they believed that it was the intention of the government to enhance their rent. Even the colonial authorities at the local level (Jamtara) were convinced that imposition of rent and its enhancement were the real issues for their uneasiness. The Santals were aggrieved that the settlement had deprived them of all those privileges which they and their forefathers had enjoyed since long.

NOTES

1. Some of the major works may be mentioned: J.C. Jha, *Kol-Insurrection in Chotanagpur* (Calcutta, 1964); *Bhumij Revolt (1832-33)* (1967); K.S. Singh, *The Dust-Strom and the Hanging Mist: A Study of Birsa Munda and His Movement in Chotanagpur, 1874-1901* (1966); K.S. Singh (ed.), *Tribal Situation in India* (Shimla, 1972); K.K. Dutta; *The Santal Insurrection of 1855-57* (Meerut, 1970); 'The Birsa Movement in Chotanagpur', in *Freedom Movement in Bihar*, Vol. I. (Patna, 1957), pp. 96-105; Ranajit Guha; *Elementary Aspects of Peasant Insurgency in Colonial India* (Delhi, 1983), K.S. Singh; 'Presidential Address to the Proceedings of the Indian History Congress', Section III, 1977, Bhubaneswar, pp. 373-97; S.P. Sinha, *Conflict and Tension in Tribal Society* (Delhi, 1993), Chap. 7 on Kherwar Movement, 1874-1942, pp. 200-23; L.S.S. Malley, *Bengal District Gazetteers, Santal Pargana* (1910) (Delhi, rpt 1984), pp. 60-151; Sumit Sarkar, *Modern Indian (1885-1947)* (Delhi, 1983), pp. 44-8; Malabika Chakrabarti, 'Alienation of Raiyati Holdings in Late Nineteenth Century Bengal', in *Rewriting Bengal's Past: Society and Culture in the Nineteenth and Twentieth Centuries,* ed. Ranjit Kumar Roy (Rabindra Bharati University, Calcutta, 1995), pp. 96-124. The number of land transactions to mahajans in Chotanagpur in the 1880s and 1890s is given on pp. 101-2. Swami Sahjanand Saraswati, *Khet Mazdur Aur Jharkhand Ke Kisan, Part II*, ed. Abadesh Pradhan (Delhi, 2001), pp. 101-204 (in Hindi). Swamiji gives an account of the plight of the agricultural labourers of Jharkhand by the landlords, mahajans, moneylenders and the colonial authorities; K.S. Singh, 'Birsa Munda and His Movement in Chotanagpur, 1874-1901', in Ravi Dayal (ed.), *We Fought Together for Freedom, Chapters from the Indian National Movement* (Delhi, 1995), pp. 27-45.
2. Irfan Habib, *Essays in Indian History: Towards a Marxist Perception* (Delhi, 1995), p. 285. Two chapters in this book on 'Processes of Accumulation in Pre-colonial and Colonial India' and 'Colonisation of the Indian Economy 1757-1900' give an excellent analysis of the working of colonialism in India.
3. Malabika Chakrabarti, 'Alienation of Raiyati Holdings in Late Nineteenth Century Bengal', in *Reviewing Bengal's Past: Society and Culture in the Nineteenth and Twentieth Centuries,* ed. Ranjit Kumar Roy, op. cit., pp. 96-103.
4. K.S. Singh, 'Presidential Address to the Proceedings of the Indian History Congress', op. cit., p. 382.
5. Sumit Sarkar, op. cit., pp. 44-5.
6. K.S. Singh, 'Birsa Munda and His Movement in Chotanagpur, 1874-1901', in Ravi Dayal (ed.), op. cit., p. 28.
7. Letter dated 8 November 1880 from Father L.O. Skresfund to the Editor, *The Statesman Friend of India,* quoted in S.P. Sinha, op. cit., p. 200.
8. Stephen Fuch, *Rebellious Prophets* (Bombay, 1965), p. 53.
9. *Bengal Judicial Proceedings* (henceforth *BJP*), August 1881, Nos. 39-40; report as to the cause of recent excitement among the Santals, No. 1076, J. Bhagalpur, 8 June 1881, from G.M. Barlow, Commissioner of Bhagalpur division and Santal pargana to Secretary to the Government of Bengal, Judicial Department (henceforth JD).

10. Letter dated 1 October 1874 from J. Boxwell, Officiating Deputy Commissioner, Santal pargana to G.N. Barlow, Officiating Commissioner, Santal pargana.
11. S.P. Sinha, op. cit., p. 205.
12. *BJP*, 1881, para 5.
13. Ibid. A note as an appendix submitted to the Commissioner by the Deputy Commissioner of Santal pargana.
14. S.P. Sinha, op. cit., p. 206.
15. *BJP*, 1881, para 12.
16. *BJP*, 1881, para 18.
17. S.P. Sinha, op. cit., p. 208.
18. *BJP*, 1881, W. Oldham, Deputy Commissioner, Dumka, 16 March 1881 to the Commissioner of Bhagalpur division and Santal pargana.

Forests, Development Ideology and Peasant/Tribal Life-World: Madras Presidency, Nineteenth and Early Twentieth Centuries

Atlury Murali

INTRODUCTION

In the 1970s through 1990s, tribes were seen as self-sufficient and isolated societies that lived in harmony with nature.[1] A significant body of environmental history concentrates on the history of state forest management and its impact on the rights of tribal people.[2] In these studies, historians have followed the argumentation which stressed on tribal communities' stable system of survival. Authors like Ramachandra Guha and Madhav Gadgil have argued that the pre-colonial societies were blessed with well-adjusted caste institutions that regulated resource use, where each community occupied a specific ecological niche in society. These ecological niches were self-contained systems of resource use that were regulated by social and cultural codes. Guha and Gadgil project the modes of resource use in tribal societies as a part of the hunting-gathering economies that include shifting cultivation. Here the economies are seen as stable and mutually exclusive closed systems that are based on prudent use of resources. They are not so harmful to ecology because their resource use is based on limited needs of small populations and highly regulative socio-cultural practices. The worship of nature, sacred groves, taboos and other practices are projected as regulatory factors in resource use.

There is also a strong historiographical tradition which articulated the argument that local society (in which include peasants and tribes) did not comprise isolated communities, for they were related to the pre-colonial state and society in various complex ways. In particular Sumit Guha's work[3] illustrates the complex interrelationship between pre-colonial regimes of natural resource management, environmental changes and tribal subsistence in the Maratha period. He also shows that the multiple occupations of tribes also yielded multiple identities and reflected the complex relationships in the Maratha political economy.[4]

None the less the dominant historiographical notion of a community is characterized by the idea of egalitarianism and homogeneity where there is little differentiation in terms of access to resources. Kinship is defined as the organizing principle of labour. The cultural aspect of society was seen as defining the way the community relate with other and also define the boundaries of the community. This is reflected in the work of Nandini Sundar and Ajay Skaria who attempt to complicate the picture by hinting at the transformation of community identities in history.[5] However this change is seen in a way that precludes any discussion of the way in which the fluidity of society in the pre-colonial period impacted upon or was determined by the production system in agriculture. This is mainly because livelihood patterns are not seen as defining the limits of either the local economy or the village community.

There are significant differences in the perspectives through which the environmental historians of the late 1980s and 1990s describe the nature of tribal life, reflecting a complex relationship between culture, ecology and subsistence. Colonialism and colonial forestry are seen the most significant factors that inflected damaging impact on the environment of the colonized society. The main effect of colonial forestry was felt in the break up of traditional village economies and stable well-adjusted systems of resource use. The forests promoted monoculture, excluded the rights of the people from forests and used the forests as a resource to extract the maximum amount of revenue and meet the goals of imperialism.

There are, however, historians like Richard Grove, who argue that the concern about climate change, soil erosion and deforestation were, articulated way before in the fifteenth century by the European thinkers, scientists and activists. He locates the British colonial environmental ideas and the state policies in the pre-colonial European history and in the process try to give a 'positive' image to the nineteenth century colonial system. For him the scientific forestry was concerned, not merely with exploitation of resources, but also with the search for a romantic ideal of tropical Eden in the east.[6]

The colonial forest policy like the other colonial projects, in particular land revenue had evolved out of many heterogeneous opinions that were expressed by officials within the colonial administrative structure. No doubt this heterogeneity of the opinion was noticed in some of the writings but the full implication of these colonial discourses is not worked out. For instance, Satpal Sangwan with the help of official correspondence published in the *Indian Forester* shows different opinions that were expressed on the notion of the state forestry. This discourse by scientists and revenue officials about scientific forestry created a context for formulation of a new system of state forestry. The colonial state orchestrated the idea that natives in a colonized society do not possess adequate scientific knowledge on conservation and therefore the state should take over the forest and manage them scientifically for the good of the society.[7] It is Ramachandra Guha who analyses this debate among the colonial officials on forest policy and broadly divided it into three categories: one category is represented by annexationists lead by Baden-Powell, who advocated that the state should take over the forests for the sake of public benefit and play the role of a benevolent landlord. Another set of officials, who were sympathizers of local communities had argued that the state role has to be passive in the forest affairs. The third category of officials however took a middle path by advocating that the state has to play an active role in protecting forests but at the same time it has to take care of people's needs, for they depended upon forests for their livelihood. Thus this discourse that the state should respect the customary rights and privileges no doubt has become the basis for the all India act, the 1878 Forest Act.[8]

Intriguingly, however, the debate in Madras Presidency had moved in somewhat opposite direction to the all India trend. In this paper we would document this Madras centred discourse with a micro level analyses of forest policy. While dealing with the impact of colonial state policies on the regulation of forests we bring into focus the issue of how the tribal and peasant communities understood the state policy and responded to it. The effort in our paper is to demonstrate the importance of the discourse that emanated from the corridors of the Land Revenue Department.

Early History

During the period 1770s and 1850s the District Records of the Madras Presidency speak of a language on the use of natural resources but without totally excluding the 'natives'. Since the colonial discourse in this period is in a formative stage the language of exclusions is not pronounced. But the policies and laws being framed were in the context of 'public utility' as synonym to Company's interests. In other words, the trade in teak is the core of the policy directive and the abundance of teak and timber, as it appears did not necessitate an elaborate policy. None the less elaborate rules were framed and enforced under the guidance of the Revenue Department.

For instance, in 1789-90, John Hamilton submitted a plan to Governor-General, proposing to promote 'the Trade in Teak Timber, by transporting it from the interior parts of the country through the zamindaris under Masulipatnam to the Coast'. The Board of Revenue was asked to express its opinion 'with regard to the abolition of the duties' levied at that time on timber articles passing through the Godavari, by different zamindars. Both the Governor-General and the Board felt that Company's trade in timber should be freed from 'vexatious enormous taxes'. One finds an interesting way of bringing in of the 'objects of public utility' into its argumentation for the abolition of taxes by the local zamindars on this 'extensive trade'. The Board ultimately abolished the taxes on timber trade in these tracts.

In regard to the Fees, Taxes, &c., which Mr. Maxwell[9] stated to amount from 60 to 65 per cent, before the timber leaves the Soubah's dominions, we gave it as our opinion that His Highness should be informed, in the event of your adopting the above recommendation of the liberal principle, upon which this Government proceeded with a view to promote the public good, and invited at the same time to a similar measure on his part for the encouragement of the traffic in question, either by an abolition of all taxes whatever on the timber passing through his territories, or else to take effective measures for preventing their increase beyond the present rate; because without the accomplishment of this latter object there was too much reason to apprehend that the good effects which might otherwise be expected from the exemption granted on the part of the Company would be rendered abortive by further exactions in His Highness' dominions, as the demand for timber increased.[10]

This language of 'liberal principle', 'view to promote the public good' by transporting timber either 'by an abolition of all taxes' or 'preventing their increase beyond the present', is to be seen in relation to their expected increase in the demand for timber. The good of the Company is the good of the people (invisible and nondescript) at the expense of zamindars' income from taxing timber trade passing through their territories. One could cite several such instances from the Godavari District Records.[11]

UNDER THE CROWN, POST-1858

The replacement of the Company with the direct British rule in 1858 appeared to have marked a change in the colonial state discourse. This change however is now based on larger interests of the metropolis rather than the interests of individuals in a trading Company. The good of the Company will have to be replaced with the good of the empire. In the Madras Presidency, this discourse was articulated in rather a longer debate between the Revenue Department and the Conservator of forests. We propose to go into these debates and see the parameters of the colonial 'development ideology' and its impact on the peasant-tribal life-world.

The first set of ideas are articulated during the 1860s and 1880s, followed by a series of policy measures that formulated the boundaries—juridical and physical—between the ecosystem and the local communities. These measures are marked by a well articulated discourse on necessity of laws and regulations as the local communities, in particular the 'illiterate peasantry' are not endowed with the necessary wisdom or moral attitude for a proper management of the forest resources. In the words of Baden-Powell:

> . . . a forest is obviously a kind of property which is peculiar, owing to the inveterate tendency of the illiterate peasantry [in all countries] to regard the forest as *un*-cultivated—as not [to their eyes] the same as an area growing corn, or apples, or cherries,—it is particularly difficult to reserve it from depredations of all kinds; people *will* think that because trees, grass, and fruits of the wild trees are the produce of nature, *therefore* it is no theft to take them. It is not felt that fold and gems are just as much the produce of nature, and that the art and skill exercised in polishing and preparing them are not distinct in kind from the care and labour that are expended on tending a forest. Special laws have to be made to protect the forests on this account.[12]

The language of exclusions and the strong presence of colonial state's interests could be seen more after the 1870s. The causation meticulously developed in its juridical discourse, the so-called 'facts' marshalled in support of its 'laws', marked a shift in the positioning of state's monopoly interests vis-à-vis the local communities, particularly the hill tribes and rural peasantry. A detailed and close reading of Brandis' *Suggestions Regarding Forest Administration*[13] would bring us closer to the realization of this process. I would not go into any such detailed analysis, as you are all very well familiar with the work in question. It is sufficient to state that the 'rules' and 'laws' governing the 'use' of forest resources are so elaborate and structured in such a language, the intelligibility itself acquires a form of coercive force and the consequent 'voluntary' alienation of the genuine and traditional users of these natural resources.

Irrigation, Commercialization and Transformation of Agrarian Economy

The root of this colonial discourse on development is the structural transformation that is at work from the 1850s itself in the Madras Presidency. The need to maximize revenue from the agrarian economy, after 1850s, brought into existence new system of water management, in particular colonial state's huge investment in big dam technology on all the major rivers in the Presidency. The most visible ones are on Godavari, Krishna and Kavery rivers. The modernization of agricultural production with new irrigational systems and transport had direct impact on the way the local peasant communities sought to use forests and *porambokes* adjacent to the villages.

In the Madras Presidency, as elsewhere in India, the practice of arable husbandry varied with soil, rainfall distribution and irrigation facilities.[14] Irrigation, in particular, was highly developed in the great deltas of Godavari, Krishna and Kavery. On the East Coast, on almost all rivers, dams were constructed from which channels had been dug to carry water to the land. Rain and drainage water was also stored in innumerable 'tanks' or artificial lakes formed by enclosing depressions and small valleys.[15]

The extent of acreage irrigated under the Krishna annicut had increased from 1,23,515 in 1859-60 to 1,99,775 in 1869-70. By 1875-6 the acreage irrigated had gone up to 2,64,256.[16] The increase in the

AREA IRRIGATED WITH DIFFERENT SOURCES

[*Figure in Thousands of Acres (omitting '000) for Madras Presidency*]

Years	Area irrigated from				Net area irrigated including other sources
	Govt. Canals	Private Canals	Tanks	Wells	
1891-2	2472	24	1698	1034	5363
1900-1	2802	25	1881	1107	5971
1901-2	2922	86	2165	1205	6585
1911-12	3535	180	3299	1442	9587

SOURCES: *Statement Exhibiting the Moral and Material Progress and Conditions of India, 1901-02 and 1911-12*, India Office, 1903 and 1913, pp. 207 and 226.

Madras Presidency, under all categories of irrigation is also impressive.

Most important among the major works were the deltaic system of the Godavari, Krishna and Kavery which together irrigated 2,712,000 acres in 1911-12.[17] In these deltas the conditions were as favourable for irrigation as in the tracts served by the great perennial canals of the Punjab in north India. The development of irrigation boosted multiple commercial crops as well as yields per acre.[18] This development of irrigation was of course most rapid under Godavari and Krishna systems in the Andhra districts. Between 1895 and 1924 the total irrigated area in the Krishna-Godavari deltas had increased by 37 per cent.[19] Details of the consequent increase in the area under first and second crops and development of irrigation district-wise are shown in the tables given below.

Between 1860s and 1900, the new irrigational systems have completely transformed the Andhra region of the Madras Presidency. The income from lands increased progressively and also the land that was brought under cultivation and commercialization. Consequently there

AREA CHARGED AS IRRIGATED

Years	Godavari system		Krishna system	
	First crop acres	Second crop acres	First crop acres	Second crop acres
1899-1900	683000	60407	588000	500
1923-4	770000	223153	757000	3000
Percentage increase	13.00	269.4	22.00	500.00

SOURCE: *Madras Resettlement Report, East and West Godavari Districts*, Madras, 1929, p. 193.

INCIDENCE OF LAND REVENUE IN WET LANDS

(per acre in Rupees)

Years	Ganjam	Vizaga-patam	Goda-vary	Kishna	Nellore	Cudda-pah	Ananta-pur	Kurnool
1864-5	2.86	5.25	3.27	4.26	5.28	6.67	3.98	6.91
1865-6	2.86	5.39	3.25	4.55	5.20	6.71	3.95	7.18
1874-5	2.89	5.19	2.59	2.26	5.31	6.44	4.86	6.30
1875-6	2.90	5.18	2.57	2.23	5.30	6.49	4.86	6.18
1884-5	3.19	5.16	2.57	2.29	5.32	6.25	4.92	6.46
1885-6	3.19	5.16	2.56	2.25	5.36	6.24	4.91	6.45
1894-5	3.23	5.51	5.82	5.78	5.44	6.19	4.86	6.33
1895-6	3.23	5.51	6.37	6.35	5.43	6.19	4.86	6.34

SOURCE: *Annual Settlement Reports, 1864-66, 1874-76, 1884-86 and 1894-96.*

LAND REVENUE INCIDENCE ON DRY LANDS

	Vizagapatam	Godavari	Krishna	Nellore	Cuddapah	Anantapur	Kurnool
1864-5	1.11	1.61	1.41	1.32	0.73	0.62	0.96
1900-1	1.01	1.58	1.45	1.13	0.77	0.65	0.93

SOURCE: *Annual Settlement Reports, 1864-66, 1874-76, 1884-86* and *1894-96.*

was a progressive increase in the land revenue incidence per acre and this increase is more pronounced in the wet cultivated lands rather than the dry regions.

By 1900 the total area that came under assessment and the increase in the assessment had an impressive growth in the almost all the districts of the Andhra under the Madras Presidency.

The increase in irrigation of cultivated lands is more rapid in the twentieth century, in particular between 1911 and 1921. This obviously had a bearing on the way the land reclamation process was done in various districts and the consequent pressure on the forests and other common property resources. But we are confining to the nineteenth century to show the effects of the irrigational modernization, for the nature of consequences and pressures are the same throughout and the only difference is the intensity that varied over a period of time.

This concentration of irrigation development in East Coast North Division also influenced the commercialization of the agrarian economy. This does not mean that the commercialization had not taken place in ceded districts or in the so-called 'dry zone'. For

TOTAL AREA AND ASSESSMENT IN 1864-5 AND 1900-1
(INCLUDING DRY & ASSESSMENT)

District	Area				Assessment			
	1864-5	1900-1	Increase	%	1864-5	1900-1	Increase	%
Ganjam	2,78,464	3,80,541	1,02,077	36.7	6,30,589	7,80,092	1,49,503	23.7
Vizagapatam	77,859	1,67,571	89,712	115.0	1,83,828	4,44,729	2,60,901	142.0
Godavari	5,77,087	9,30,924	3,53,837	61.3	16,94,375	36,72,145	19,77,770	116.7
Krishna	16,09,423	20,25,814	4,16,391	25.9	29,12,674	48,67,869	19,55,195	67.0
Nellore	6,94,155	10,74,236	3,80,081	54.8	16,34,924	21,06,810	4,71,886	29.0
Cuddapah	12,49,908	13,12,063	62,155	5.0	17,73,802	17,05,097	-68,605	-3.9
Anantapur	24,93,709	27,85,573	2,91,864	11.7	20,99,296	23,76,814	2,77,538	13.3
Kurnool	11,62,239	12,89,971	1,27,732	11.0	13,18,812	13,56,800	37,988	2.7

SOURCE: *Annual Settlement Reports, 1864-65* and *1900-01.*

DEVELOPMENT OF IRRIGATION IN DIFFERENT ANDHRA DISTRICTS

Name of division and district	1911			1921			1931		
	% of cultivable area			% of cultivable area			% of cultivable area		
	Net cultivated	Double cropped	% of cultivable area which is irrigated	Net cultivated	Double cropped	% of cultivated area which is irrigated	Net cultivated	Double cropped	% of cultivated area which is irrigated
Madras Province	65.3	8.6	25.9	63.4	8.3	28.3	60.4	8.1	28.6
Agency	33.8	1.1	40.7	41.8	0.6	26.6	38.1	0.8	6.7
Deccan	74.1	3.4	7.7	66.5	2.4	7.5	65.3	2.3	8.9
Cuddapah	62.1	7.1	19.9	53.9	5.2	20.1	51.5	4.3	23.5
Kurnool	78.1	2.7	4.7	72.0	2.1	4.6	71.4	2.0	5.7
Ballary	85.8	1.5	2.7	80.4	1.0	2.5	78.2	1.6	2.9
Anantapur	65.4	4.0	10.2	56.0	2.4	10.3	56.1	2.2	11.7
Chittoor	52.6	9.9	47.6	40.2	7.7	43.4	37.6	7.7	38.6
East Coast North	65.9	12.7	37.3	64.3	13.1	43.3	62.8	12.5	40.0
Ganjam	81.0	16.5	43.5	75.5	14.6	50.5	78.8	14.6	46.8
Vizagapatam	62.7	23.9	47.7	59.9	22.0	49.5	62.1	22.7	38.5
Godavari	70.4	21.3	49.6	69.3	22.6	69.7	71.7	20.5	50.4
							(61.6)	(15.1)	(67.6)
Krishna	62.4	8.9	45.5	63.6	12.5	55.6	66.1	8.6	44.3
Guntur	78.8	11.6	13.5	76.4	11.0	16.2	73.8	10.3	15.5
Nellore	49.7	5.6	36.7	46.4	4.8	37.3	42.3	4.9	42.9

NOTES: Chittoor was in East Coast Central Division. Figures in ( ) are for west Godavari district and others are for east Godavari district.

SOURCES: *Census of India, 1911* (Madras), Vol. XII, Pt I, p. 16; 1921, Vol. XIII, Pt I, p. 27; *Season and Crop Report of the Madras Presidency, 1930-31*.

example, the shifting of acreage from food crops to industrial crops did develop in both the divisions, though with a degree of difference, due to the conditions of production. In coastal Andhra, rice emerged as the dominant commercial crop, while it was the industrial crops that increased substantially in the ceded districts. Unlike the ceded districts the delta districts were blessed with rich alluvial soil and developed irrigational facilities. Consequently, they were well suited for rice cultivation. On the other hand, due to the nature of soil and lack of proper irrigational facilities the ceded districts opted for industrial crops which were suited to those particular seasonal conditions.[20] Thus, both the regions were drawn into the network of commodity production and integrated with the international colonial market and trade. As A.K. Bagchi observes that 'one difference between the relatively fertile and easily accessible regions and the "dry" regions of interior was that it was easier to centralize the surplus in an exportable form in the former areas; hence they provided a lusher field of operation for European traders and capitalists'.[21] Perhaps, this difference will not appear so important if one sees it within the overall framework of colonial exploitation.

In the 'dry' region, i.e. the Deccan districts, the phenomenal growth of cotton and groundnut, which constituted the bulk of exported industrial crops, was striking. Cotton production increased by 18.6 per cent whereas groundnut increased by nearly 306 per cent.[22] On the other hand, the increase in indigo and castor was concentrated in the East Coast North, the percentages of increase being 109.8 and 10.8 per cent respectively. Of all the industrial crops in the coastal districts it was tobacco which increased steadily over a period of time. In 1901-2 the area under tobacco was 46,760 acres. This increased to 91,989 acres by 1910-11 and 1,46,562 acres by 1931-2.[23] Numerically, these increases may look meagre but it was quite sizeable if seen in the light of the nature of soil in these two divisions. As a whole, it can be safely suggested that while the development of paddy and some of the industrial crops was mostly concentrated in East Coast North districts, the ceded districts emerged as dominant producers of important industrial crops like cotton and groundnut. Thus both divisions were affected by the commercialization of agriculture and consequent emergence of new agrarian classes under British rule, especially in the 1910s and 1920s.

However, in spite of impressive development in the ceded districts, the area under industrial crops in the Andhra districts as a whole (11 districts) was nevertheless small. Only about 13 per cent of the crops grown were industrial crops during 1910-11;[24] these figure progress to only a little over 14 per cent in 1920-1.[25] These crops were in the main composed of cotton and oil seeds in equal parts;[26] and items like sugar and dyes increased from less than one per cent in 1910-11,[27] to only 1 per cent in 1920-1.[28]

The development of these commercial crops in almost all regions in the Madras Presidency was mostly influenced by the demand in national and international markets. For instance, it was reported by the district administration that in Vizagapatam, the 'jute of the district locally known as "erragogu" has risen into much prominence in recent years owing to the increase in the price of that fiber due to demand from foreign countries. It is chiefly grown in red soil in the center of the District round Bobbili, Salur, Gajapatinagaram, etc.'[29] To quote another example, the area under indigo had been steadily declining till there were only some 80,000 acres under it in 1910-11 due to lack of international demand. However, it increased to over 3,00,000 acres in 1917-18 owing to the scarcity of synthetic dyes during the First World War but again fell to 1,09,298 acres in 1920-1.[30] As an example we have given here in a tabular form the indigo crop statistics to demonstrate these wide fluctuations.

By 1900, the transformation of the Presidency by the new irrigational systems was also effected by better communication network, both internally and externally. In the case of Godavari district, it was recorded that 'the district practically dividing the two portions. The Godavari delta is about 2,000 sq. miles, which is traversed by 500 miles of navigable canals, in other words, there is .1 mile of canal for every 4 sq. miles. There are 255 miles of metalled road in the delta or 1 mile of road for every 8 sq. miles.'[31] Since the railway rates were considerably higher the canals, which ran parallel to the railway line from Bezawada to Cocanada, frequently carried large quantities of delta produce and enormous passenger traffic. Therefore, it was proposed to take up the construction of more village roads to connect the village with the canals. In the first two decades of the twentieth century that many new 'railed roads' were constructed in the upland taluks, connecting it with the rest of the presidency. The main line

AREA UNDER INDIGO IN THE MADRAS PRESIDENCY
BETWEEN 1890-1 AND 1924-5

Year	Area in acres	Year	Area in acres	Year	Area in acres	
					Govt.	Zamindari
1890-1	255511	1907-08	141876	1900-01	252054	7408
1891-2	212255	1908-09	79898	1901-02	239387	7667
1892-3	301201	1909-10	94750	1902-03	212238	8649
1893-4	442439	1910-11	82562	1903-04	241917	19631
1894-5	527081	1911-12	90324	1904-05	126098	6258
1895-6	486938	1912-13	66552	1905-06	113858	1401
1896-7	454748	1913-14	55246	1906-07	180723	2762
1897-8	323931	1914-15	71683			
1898-9	210610	1915-16	221974			
1899-1900	276210	1916-17	459066			
1900-1	254462	1917-18	324385			
1901-2	247054	1918-19	144357			
1902-3	220887	1919-20	101300			
1903-4	261548	1920-21	112246			
1904-5	132356	1921-22	197282			
1905-6	115259	1922-23	141316			
1906-7	183485	1923-24	91090			
		1924-25	57600			

SOURCE: GOM, Development Department, G.O. No. 646 (Mis.) dated 1 May 1925, pp. 10-11.

completed in this period was the 'Light Railway' from Samalkota to Yellesavaram via Peddapuram and Jaggampeta, which was taken across country connecting the sugar company's factory at Samalkota.[32]

By 1900 the Madras-Calcutta main broad-gauge line linked the Krishna district with the other East Coast districts. Metre-gauge lines had already connected the Newada main junction with the interior Krishna centres. There were six important feeder lines: (1) Masulipatam-Gudivada-Bezawada of 50 miles metre gauge; (2) Narasaraopet-Nakarikallu-Pidugurala-Dachepalle-Gurajala of 42½ miles; (3) Guntur-Narakoduru-Chebrolu-Ponnur of 19 miles; (4) Narakoduru-Tenali Jampani-Kolluru of 22½ miles; (5) Gurajala-Rentachintala-Macherla of 15 miles and (6) Jampani-Bhattiprolu-Repalli of 15 miles. These railway lines tapped the 'rich rice-producing country (villages)' and 'the cotton and dry grain growing tracts' of interior Krishna.[33] 'The remainder of the delta (was) already', it was observed in 1899, 'so well supplied with navigable canals that no railed road probably could compete with them'.[34]

As early as 1893 another line was opened integrating the Cocanada and Vizagapatam with the main East-Coast line. Apart from the above route, the Bengal-Nagapur railway line had linked Ganjam and Vizagapatam districts with Bengal and Orissa. In Vizagapatam the chief trade route was however the road from Bhimilipatam through Vizianagaram and Salur to Jeypore. Many interior centres were linked with the Vizianagaram-Raipur railway line which passed through Madgole to Anakapalli and Parvatipur via Palkonda to Chicacole Road railway station. This communication tapped the hill trade and the Central Provinces trade at Parvatipur and heavy trade traffic of Anakapalli and Palkonda and the products traffic from the Government forests in Ganjam and Vizagapatam.[35]

Thus, by the turn of the twentieth century communications through railways, canal navigation and roads had provided an efficient and cheap means of transportation in most of Andhra. No doubt, by 1910-20, the improvised railway feeder networks and the metalled roads for motor transport practically eclipsed the canal navigation,[36] but the emergence of other means of communication widened the scope for transport of commodities from different local market zones to the national and international trading ports as well as of the European goods into the interior Andhra villages. In other words, with the expansion of communications there emerged coherent trading zones both localized and centralized, which integrated the Andhra agrarian economy with the national and the colonial world market and thereby influenced the production conditions in agriculture. The interior was opened up to the coast and through the coast to the international market structure. It has been rightly observed in a recent study that during the twentieth century the Andhra region 'had different levels of market structures and the expanding transport and communication system uniformly integrated the regional market'. The integration not only functioned as the medium of quick commodity circulation at the local and regional level but also contributed a necessary amount in the regions widening integration with the external market. Consequently, the development of the regional market increasingly drew the commodities from the interior to the national and international export.

This tremendous improvement in communications and the internal and external trade enabled the rich peasants, merchants and traders to control the regional as well as the national markets. They were able to import or export food grains from one district to another or from one province to another at times of scarcity, and thereby make huge profits.[37] They not only monopolized the trade at the local, provincial and the national level but also controlled the marketing conditions and the price levels to their advantage.

In fact, the very process of modernization and commercialization of agrarian economy with well integrated network for the marketing of the produce had its peculiar pressure on the use of the resources like forests and common lands. The colonial discourses on the nature of state control that has to be developed had been in response to the pressure of modernizing rural agrarian economy. The diversity of opinions reflected in the land revenue department and the forest conservation department could be contextualized only in relation to the above modernization process of agrarian economy. For the post-1858 colonial state's interest was maximization of revenue from the agrarian economy, a move away from the early overriding concern with the extraction of teak and timber for shipbuilding and railways. This change could be seen more clearly in the arguments of collectors of various districts from 1870s.

MODERNIZATION AND FOREST CONSERVATION: THE STORY OF MULTIPLE VOICES

Coming back to the main story, the colonial state's effort to manage the forests and the response of the district collectors, from the beginning was not unified in nature. There were two factors at work: one was the overriding concern of the Revenue Department to maximize the revenue collection by retaining some form of juridical control over the management of the forests, in particular the ones adjacent to the villages and the other was the Forest Department that constantly sought to take over the complete management of the forests more in the name of 'scientific' and 'modern' management. Responding on the various provisions of the Draft Bill proposed for the management and preservation of forests in 1871, the Collector of Madura wrote to the Acting Secretary to the Board of Revenue that summed up Revenue Department's perspective. Recalling from his first-hand knowledge of the operation of the forest conservancy and the scant respect shown by the forest officers for the communal rights of the people, he wore that 'I am sorry to say that the former has been a complete failure and that the latter have been utterly neglected.'

About the year 1858, recalled the Collector, D. Arbuthnott that the Government became alarmed at reports from many districts of the rapid decrease in the extent of forests, and the Forest Department was hastily organized to put up a stop to further deterioration. No law was passed, but rules were framed under the orders of Government. Under these rules no forest was accessible to the people unless on licenses either paid for or free. The rules unfortunately fall most heavily on those who are at a distance from the forests and who are not allowed even firewood without paying for a license, so that the unfortunates who have no forest within the limits of their village have not only to bear the cost of the carriage of firewood, but also to pay for their licenses, whilst those who have a forest at their doors escape both cost of carriage and of license. This inequality does not apply to fuel alone, but to every other article for which wood is required, such as ploughs, rafters for houses, etc.

The decrease in the area of forests has arisen solely from one cause, viz., the extension of cultivation. The idea that forests have suffered seriously from indiscriminate felling and wanton mischief is simply absurd. Felling timber is a labourious occupation and not very remunerative, and these considerations alone are sufficient to prevent mere wanton destruction.

In this district almost the whole of the Pereacolum Talook was, at the beginning of this century, one vast jungle. People are now alive who remember all the Punjah land covered with forest, but with rare exception the forest consisted entirely of the common Babool, which, in this part of the country, invariably assumes the umbrella shape. The extirpation of such a forest was only an unmitigated good. Fever disappeared and the land is now largely cultivated. That the forests of the district are amply sufficient for the wants of the people I am perfectly satisfied.

From the examination of the forests I have made, and from the statements of many persons in the district, I am also of opinion that no considerable quantity of the more valuable kinds of wood, such as Teak, Vengai, and Blackwood ever existed in this district. The idea that our jungles have deteriorated rapidly and to such an extent as to cause alarm for the welfare of the people is totally groundless. There is sufficient forest for all the wants of the poorer classes for many thousands of years. The richer classes can take care of themselves.

The result of the proceedings of the forest Conservancy Department is that they have made themselves universally hateful to the people. This is not the fault of the European Officers but of the system. . . .

Under the present system, as far as my personal knowledge extends, the jungles are not much worse, but not an iota better than when the forests were allowed to take care of themselves. The State has gained no income, the people have not only been heavily taxed, but also greatly inconvenienced and annoyed, and the forests are in the same state they were ten years ago. In most districts I believe there is no occasion for a Forest Department at all; with a very small establishment the forests could be managed by the Revenue Department, and, if not better, the management would certainly be more economical and far less harassing to the people.[38]

Most of the Collectors in the Madras Presidency echoed similar observations basically motivated by the needs of the plain peasants' needs rather than the needs of the tribal communities living in the forests. Whether it is the issue of access to the firewood, wood for agricultural implements and land for reclamation [cultivation] or the need for fodder, the argument of the Land Revenue Department was more tuned to the needs of the peasants, particularly the poor peasants. For they have rightly observed that the rich peasants are capable of taking care of themselves as they could get the necessary permits with payment to the government. The developmental ideology was more tuned towards the needs of the agrarian economy that is, from the 1860s getting transformed and commercialized under the impact of new irrigation systems, transport facilities and marketing systems. The argument of 'community' rights and privileges is slanted towards the needs of the agrarian economy.

None of the Collectors, however, were against the demarcation of the state forests in the zones that have dense forests, but were opposed to the incorporation of 'scrub jungles near villages' into the definition of forest, for such a system of defining forests would bring 'enforcement of the Forest rules by forest subordinates' into the villages and that 'would be oppressive in the extreme'.[39] 'The forests of this district,' wrote Chase, the Collector of Kurnool 'and I may say of this Presidency, are all forests in which the right of Government is limited by the customary privileges of village communities and hill tribes.' To quote further:

In this district many of these privileges, particularly the pasturing of cattle, the gathering of leaves for manure, and the cutting of firewood for domestic purposes and for sale, have been enjoyed from time immemorial both by the villages in the immediate neighbourhood of the forests, and also by all the villages throughout the district.

It has also been customary for the ryots of Nellore and Kistna Districts to graze their cattle for six months of the year on the eastern slopes of the

Nallamala hills, and though their right is only permissive and acquired by long usage, the closing of forest pastures against them would tell seriously upon the breeding of cattle and the agricultural prosperity of these districts and affect the supply of cattle to the Kurnool District, for which it almost entirely depends on the Coast Districts.

Commenting on the Part 4 of the proposed Forest Act, the Collector observed that it:

. . . provides for the adjustment of private rights, and as a ground work of this settlement Section 18 lays down that no fresh rights shall accrue. This will operate against the extension of cultivation. Under the present ryotwari system, lands not required by resident villagers are offered to strangers, and if this offer is to be saddled with a condition that the stranger must pay a grazing fee for his cattle in addition to the land assessment while others pay the assessment only, it will no be accepted and the extension of cultivation will be impeded. In like manner the inhabitants of the village who undertake cultivation for the first time, or who may increase their holdings after the demarcation of the forest, will be compelled to pay a similar fee, and in this way the old system of Poollery will be revived with all its attendant evils.

Similarly the Collector's observations on the Sections 24 and 25 that proposed to give the Conservator the power of closing forests argued that:

In many cases this will operate as deprivation of admitted forest rights. If for instance the portion of Nallamala between the Nandikanama pass and Mantralakanama pass were closed, the people of Nundial and other villages who chiefly cultivate wet lands (and this cultivation may increase under the new Irrigation Canal) will have to go some 30 or 40 miles for leaves for manure and for pasture lands for their cattle and a great hardship will be inflicted by Section 26 on hundreds of helpless women whose only means of livelihood for several months of the year consists in gathering for sale jungle produce, such as Kaliva (Carissa Diffusa), Jana (Grevia Rotundiflora), and other fruits, and various kinds of leaves eaten by the natives. During the famine of 1866, in the talooks of Markapur, Cumbum, and in parts of Nundial, Sirwell, and Nandikotkur, thousands of people lived on Boddikura (Rivea Hypocrateriformis) and other herbs, and the hill grass supplies food for cattle which suffered much for want of fodder. Inconvenience will also be felt by native doctors, most of whose medicines are found in the forests, probably at the very season when the forests may be closed. When the forests are open, inhabitants of other than the favoured villages, and who may be the only ones who possess a knowledge of medicine, will not be able to collect forest produce, as they will have no forest rights. And those who have forest rights are prevented by Section 26 from selling the produce of the forests. I do not think that the Conservator's Department will make any arrangements for the gathering of the medicines for public sale.

Section 26 prohibits the alienation of forest rights by sale or otherwise, while Section 18 disallows new rights. These provisions will tell seriously on the value of landed property, and render it almost unsaleable; few would buy lands and cattle unless allowed with other ryots the means of feeding cattle and procuring the usual vegetable manure. In the latter part of this Section, Courts of Justice are authorized to sell forest rights. This seems somewhat inconsistent, because the Courts only act on behalf of the owner and cannot exercise privileges which the owner himself does not possess.

The Bill is silent regarding the hill tribes scattered throughout the forests, and who live almost entirely by the sale of hill produce. This right it is proposed to ignore by Section 25, and compensation will not be made for its extinction by Section 27. I do not think it possible to close any forests or forest pathways against the Chensus. They will make their own way, and come down to see their customers at the foot of the hills, and if they are allowed egress and ingress, it is not practicable to prevent others from taking advantage of their routs. Any attempt to regulate the movements of Chensus must prove a failure and endanger the safety of travelers.[40]

Though touched upon, what is not clearly spelled out elaborated in the above observations of the Collectors is the effect of the expansion of wet cultivation by the canal irrigation on the grazing and pasture lands, for they were reduced drastically. This seriously affected the condition of the cattle, 'which represent the only tangible wealth of the agricultural classes, deprived the ryot of the facilities of manure, and resulted in a general deterioration of the productive capacity of the soil'.[41] Prior to the construction of the annicut, every village had grazing ground and pasture. After its construction, ryots were forced to send their cattle to the distant upland areas, which resulted in the loss of manure. During the agricultural season they could not send their ploughing cattle. The Collector of Godavari district, Henry Morris said that 'the want of good pasturage was undoubtedly the cause of the great mortality of cattle in the Godavary district'. In 1863, he said:

There is very little grazing land in the Delta taluqs, as almost the whole of the soil has been taken up for cultivation, and the dry land converted into wet. The cattle are generally sent to the upper taluqs for pasture during the rainy season, and brought back again when the rains are over.[42]

Further, Henry Morris said that the grass grown on land watered from the annicut 'causes the cattle that feed on it to contract a disease, called in Telugu *Jalagarogam* which generally proves fatal' and 'inflammation of the throat is also another dangerous disease to which the cattle are peculiarly liable'.[43] This disease prevailed in the delta taluqs only and was unknown before the construction of the annicut and its numerous canals. The disease affected the cattle by feeding on the grass which grew luxuriantly on the banks of the Channels, tanks, and swamps that were filled with the water from the Godavari river.[44] Therefore, farmers failed to cultivate properly their lands due to unhealthy cattle, lack of adequate manure and less pasture land.

This process repeatedly brought into discussion the need to revive the customary management of forests and the passing on the control over the management of forests, adjacent to the villages and agricultural zones into the hands of local community. The concept of 'village forests' repeatedly raised its head not only in some of the officials' correspondence but also in petitions submitted by the peasants.

The strong indictment of the forest administration and its rules by the Revenue Department though temporary put off the forest bill, the lobby for a strict and comprehensive forest act continued to operate. This, in 1875, gave birth to a memorandum on the forest legislation by D. Brandis. It is the views of Brandis and his strong supporter, Baden-Powell[45] that ultimately reflected in the first comprehensive forest act of 1882. This act in practice settled the conflict by giving primacy to the forest administration, though in certain areas the Revenue Department was given powers to sanction access to the peasants to fodder, firewood and *porambokes*. With the system of working plans, there came into operation a well organized Forest Department:

> Where no working plans exist and until such plans have been made, the annual plan of operations must be based on the general principles of forest conservancy. These annual plans will prescribe what work is to be done in the demarcation of boundaries, road-making and other communications, fire-protection, closing forests against grazing, cultural operations, thinning and cutting, and will generally indicate all work that should be done during the year in each forest range of the district.[46]

The 1882 forest act, by bringing in strict forest control over the forests and grazing lands, shifted the conflict between two government departments, for there arose a serious opposition to the act from the peasant and tribal communities. The forest officers' powers, the policy of forest reservation and settlement trampling on the traditional rights of local communities and the vexed question of 'fodder' continue to dominate the post-1882 act in the Madras Presidency.

This however does not mean that after passing of the Forest Act, the nature of opposition from the peasants did not change. Once the state translated its discourse of conservation of forests into a legal act, the peasant proprietors in the villages also started re-articulating their demands in juridical terms and customary rights having been supported by an elaborate juridical discourse. Let us illustrate this point. It was reported in 1862 that the ryots of Bundapalle village in Chittoor taluq have been damaging the bamboo bushes and trees and grazing in the Naragani Konda jungles. This violation was done by the villagers led by Subba Reddi of Chittapara village in the above *taluq* in the name of asserting their 'ownership' over the forest that was initially sought to be regulated but by the middle of 1888 formed and notified into a reserve by the forest department. Here the violation of the regulation in the name of asserting their traditional ownership over the forest turned into a plea for access on payment or permit system once it was reserved by the government.[47] This transformation is typical of any agrarian community which enjoys legal ownership over land. They would not violate any law, lock stock and barrel for such a violation go against the principle of private property introduced in land by the colonial state which helped consolidation of peasant on land.

It is in this changed contest that the colonial state also tried to use a policy of conciliation to the plain peasant proprietors with small and regulated concessions executed by the Revenue Department. In the case of tribes however the state had followed a different policy. First, the locally mediating elite were brought into the colonial rule of law and its logic of development. Any movements opposed to colonial regimes of coercive development were countered ideologically and suppressed militarily. Let me illustrate this process in relation to Gudem and Rampa regions in Andhra. But then what is the colonial development ideology?

In the Visakhapatnam District Collectorate Records we found an interesting typed 'Speech of Agent to Governor, Visakhapatnam district in Korukonda, Addressed to Muttadars on 23 November 1918', the Agent was H.A.B. Vernon. The occasion was the sanctioning of Sanads to the muttadars by which they were entitled to hold to their muttahs.

I will have the Sanads read to you and will then explain to you certain important clauses in them . . .

A. I will draw your attention first to the rule as regards forests. As you have been often told on previous durbars held by myself and by my predecessors and by your Assistant Agents, the Government attaches the greatest importance to the preservation of forests. Not only do these forests contain timber and other produce which are of great value but also they serve to collect the moisture which is the source of water for the rivers and streams which flow from these hills into the plains. Destruction of forests by podu and by fire will not only diminish the water supply in the rivers and will also cause these rivers to be filled with sand and silt which is carried down to the plains and fills up the channels and tanks which serve as irrigation sources for the crops of the ryots in the plains. Government is firmly determined that these forests shall be preserved, and I once more remind you of your duty to assist in their preservation.

B. The next point in the Sanad, which I wish to emphasize, is that you should endeavour to extend the permanent cultivation of the level tracts in your muttahs and to encourage ryots to come here and to take up permanent cultivation. In order to make this easier Government are allotting every year large sums of money for the improvements of roads in the Agency. I hope that in a few years you will have an important road running from below the ghats to Lammasinghi and from thence to Malkanagiri. As soon as this road is open for cart traffic you will find a better market for your produce and you will be able to induce setters from the plains to come here and to take up permanent cultivation. The shandy at Kondasantha to which most of you take your produce will then be moved to Lammasinghi at the top of the ghat, and the influx of carts will enable you to get better prices for your grain and other produce. It is therefore your duty to assist in every possible way and especially by the provision of labour, the efforts of government to construct this road and all other roads in the Agency.

C. The next point to which I would call your attention is that you are forbidden to alienate the lands of your muttah by mortgage, sale, gift or otherwise. If you follow the wishes of Government and development the cultivable land in your muttahs there will be no need for you to take loans or to mortgage any part of your lands, which will become more valuable every year if properly developed.

D. I next call your attention to the fact that Government does not recognise any hereditary right to the office of Muttadar. You hold your muttah during the pleasure of Government and as long as you fulfill strictly and faithfully the conditions set forth in this Sanad, Government will retain you in your office. But if you do not so fulfill those conditions Government will not hesitate to remove you from your office.

E. In clause V you are empowered to punish any of the Munsifs within the limit of your muttah. It is your duty to see that the Munsifs carry out your orders and the orders of the Agent to the Governor and his Assistants and if they do not do so you are hereby empowered to punish them.

F. In future in all cases in which Muttadars are warned, punished or suspended by the Agent to the Governor for any breach of the conditions of their Sanad or other misconduct an endorsement stating the nature of such punishment will be made in the Sanad and on the other hand in cases where the conduct of a Muttadar has been especially meritorious either from the successful development of his muttah or on account of any special services rendered to Government, an endorsement in red ink will be made in the Sanad in order to form a permanent record of the approbation of Government.

Since the previous Darbar, the Agent read out several cases where the government has either suspended or warned for not carrying out the duties set in the Sanad and also for 'allowing podu cultivation' in their muttahs. For instance, Gaduturu Pandu Padal, the muttadar of Ginelakota 'was suspended for I year for deliberately allowing podu with the full knowledge of the Agent's orders'.[48] Similarly the muttadar of Sujankota, Mattam Pothanna Padal was suspended for one year 'for not attending to Government duties properly and for allowing a large number of podu in the muttah and not collecting assessments charged thereon'.[49] And the Sanad of Matam Dappu Padal of Rudakota muttah contained a warning: 'The Muttadar allowed cultivation of old podu in the muttah: he was warned that even this should not be allowed without the Agent's permission and that he would be removed if he allowed any more podu.'[50] Several muttadars were even fined for allowing podu in their muttahs and also warned for not giving 'proper assistance to the Special Assistance Agent when required to do so for work on the Agency road'.[51]

CONCLUSION

In this paper we tried to show in brief the evolving relationship between the forests, colonial paradigm of conservation dictated by the Western notion of development and the natives in what is generally characterized the peasant/tribal life-world. In this history what is crucial is the internal dynamics of change: modernization and commercialization of agrarian economy rather than the influence of pre-1858 colonial concern with timber and teak. This paper only aims at hinting at the importance of the study of agrarian economy's influence on the forests and policies that aimed at conservation, so that a detailed study could be undertaken at regional and national level to highlight this dimension of environmental history of colonial India.

NOTES

1. See Christopher Von Furer Haimendorf, *Tribes in India* (OUP, Delhi, 1982); K.S. Singh, *Birsa Munda and His Movement* (OUP, Delhi, 1983).
2. Ramachandra Guha, *The Unquiet Woods: Ecological Change and Peasant Resistance in the Himalaya* (OUP, Delhi, 1989); K. Sivaramakrishnan, *Modern Forests* (OUP, Delhi, 1999); Ramachandra Guha and David Arnold (eds.), *Nature, Culture, Imperialism* (OUP, Delhi, 1995); Ramachandra Guha and Madhav Gadgil, *This Fissured Land: An Ecological History of India* (OUP, Delhi, 1992); Richard Grove, V. Damodaran and S. Sangwan (eds.), *Nature and the Orient* (OUP, Delhi, 1998).
3. Sumit Guha, *Environment and Ethnicity in India, 1200-1991* (CUP, Indian edn. Foundation Books, Delhi, 1999).
4. Sumit Guha, *Environment and Ethnicity*, pp. 55-61.
5. Nandini Sundar, *Subalterns and Sovereigns: An Anthropological History of Bastar, 1854-1996* (OUP, Delhi, 1997); Ajay Skaria, *Hybrid Histories: Forests and Wilderness in Western India* (OUP, Delhi, 1999).
6. Richard Grove, *Green Imperialism: Colonial Expansion, Tropical Island Edens and the Origins of Environmentalism* (CUP, Cambridge, 1995, Indian reprint OUP, Delhi) and *Ecology, Climate and Empire: The Indian Legacy in Global Environmental History, 1400-1940* (OUP, Delhi, 1998).
7. Satpal Sangwan, 'Making of popular debate: the Indian forest and emerging agenda of the state forestry in India, 1875-1904', *Indian Economic and Social History Review*, Vol. 36, No. 6, 1999.
8. Ramachandra Guha, 'An Early Environmental Debate: The Making of the 1878 Forest Act', *Indian Economic and Social History Review*, Vol. 9, No. 1, Jan.-June 1993, pp. 119-50. Also see Mahesh Rangarajan, 'Imperial Agenda and Indian Forests: The Early History of Indian Forestry', *IESHR*, Vol. 37, No. 2, 1994.
9. Mr. Maxwell from his long residence in the Northern Circars having acquired an intimate knowledge of this subject was requested to report therein for the information of the Board, and the observations used by the Board were from his Minute.

10. The above information is from *General Reports of Board of Revenue, 1786-1794*, Madras, 1871, Vol. 2, [1789], pp. 80-1.
11. For instance, see *Godavary District Records*, Vols. 937 and 4638.
12. B.H. Baden-Powell, *Memorandum On Forest Settlements In India*, Calcutta, 1892, pp. 1-2. Emphasis is in the original.
13. D. Brandis, *Suggestions Regarding Forest Administration in the Madras Presidency, dated Madras, 10th January 1883* (Government Press, Madras, 1883).
14. *A Statistical Atlas of the Madras Presidency, 1910-11* (Madras, 1913), p. 5.
15. The detailed story of irrigation is reconstructed from *Reports On The Direct And Indirect Effects Of The Godavery And Krishna Annicuts, In Rajahmundry, Masulipatnam, Guntoor, &c. And The Coleroon Annicuts, In Tanjore, And South Arcot* (Fort Saint George Gazette Press, Madras, 1858); Proceedings of the Board of Revenue (hereafter PBOR), Fort Saint George, 22 May 1858, pp. 549-53 and No. 4581, 10 Dec. 1858, pp. 1648-56; *Administration Report of the Public Works Department Irrigation Branch for the year 1920-21 and 1938-39*, Madras, 1922 and 1940, especially statements showing the financial results of irrigation under productive and unproductive systems in Andhra districts.
16. PBOR, No. 1229, 1878.
17. *Moral and Material Progress, 1911-12*, p. 322. Perhaps the best example would be Vizagapatam district to show the link between the development of irrigation works and the increasing diversification of the cropping patterns. See General R. Dis. No. 1682/28c2 dt. 6 May 1928 (Vizagapatam Collectorate Records).
18. For instance see return of the yield per acre of the principal crops cultivated in India (1892), pp. 1-3: the link between the proper water facilities and the yield on lands was shown in a case study. *Report on the Operations of the Department of Agriculture, Madras Presidency, for the official year 1920-21* (Madras, 1922).
19. G.O.M., Board of Rev. G.O. No. 29 (Land Rev. & Sett.) dated 18 May 1927, pp. 36-8: It was also shown by Dharam Narain that the average under paddy had increased due to the expansion of irrigation. Dharam Narain, *Impact of Price Movements on Areas under Selected Crops in India, 1900-1939* (CUP, Cambridge 1965), p. 127.
20. See note No. 151 and G.O. No. 1171, Revenue, dated 26 Nov. 1906: Board of Revenue (Rev. Sett. Survey, Land Rev. & Agriculture), No. 46, dated 6 Feb. 1909; A.V. Ramana Rao, *Economic Development of Andhra Pradesh 1766-1857* (Bombay, 1958), p. 283; V.V. Sayana, *The Agrarian Problems of Madras Province* (Madras, 1949), pp. 3-5.
21. Amiya Kumar Bagchi, 'Needed Political Economy of British South India' (Review article of Baker and Washbrook books) in *Social Scientist*, Aug.-Sept. 1978, p. 99.
22. Percentage of area under cotton and oil seeds to total area cropped

Year	Cuddapah	Kurnool	Bellary	Anantapur
1900-3	12.60	16.48	20.04	16.41
1912-14	20.46	22.29	25.25	22.75
1919-20	22.74	21.72	25.68	19.51
1931-2	22.42	29.89	36.60	26.25

SOURCE: *Season and Crop Reports of the Madras Presidency for the specified years.*

23. *Season and Crop Report of the Madras Presidency for the Years 1901-2, 1910-11 and 1931-2.*
24. *A Statistical Atlas, 1910-11*, p. 6.
25. *A Statistical Atlas of the Madras Presidency, 1920-21* (Madras, 1923), p. 6.
26. Ibid., *1910-11*, p. 6 and *1920-21*, p. 6.
27. Ibid., *1910-11*, p. 6.
28. Ibid., *1920-21*, p. 6.
29. Gen. R. Dis., No. 6273/21, 83 dt. 14 Aug. 1922 (Vizagapatam District Collectorate Records), p. 17 (hereafter VCR).
30. *A Statistical Atlas, 1910-11*, p. 6; *1921*, p. 6; to quote another example it was observed that in Vizagapatam district, 'Ground-nut which was originally confined mainly to Chipurpalli, Palkonda (rich rice growing area) and two or three other taluks, is now more or less cultivated all over the district. Very popular with the ryot as it has been paying him well.' General R. Dis. No. 6273/2183 dated 14 Aug. 1922, p. 17 (VCR).
31. Memorandum by the Members of the Sub-Committee of the Dt. Bd., Godavari, in R. Dis. 235, dt. 21 Aug. 1900, Gen. (VCR), enclosure No. 3.18.
32. Letter R. Dis. No. 112/D.B. dt. 30 Dec. 1899 from A.E. Castlestuart, Collector and Pres. Dt. Bd., Godavari; letter dt. 30 Nov. 1899 from the Chairman, Chamber of Commerce, Cocanada, to the Collector of Cocanada, enclosure No. 1; Letter from J. Innes, Chairman of Chamber of Commerce, Cocanada dt. 8 Aug. 1895, to the Collector, enclosure No. 4 in ibid., pp. 16-20; Resettlement Report of Krishna, West and East Godavari districts, in Board of Rev. (Land Rev. & Sett.) No. 29 dt.18 May 1927, pp. 30-1.
33. Letter Dis. No. 1868/1 dt. 18 Dec. 1899, from R. Morris pres. Dt. Bd., Krishna, letter dt. 10 Dec. 1899 from the Local found dt. Engineer, Krishna, enclosure No. 1: letter Dis. No. 1932, Gen. Dt. 15 Dec. 1899 from R. Morris in R. Dis. No. 235, dt. 21 Aug. 1900, pp. 20-5.
34. Letter Dis. No. 1932, Gen. dt. 15 Dec. 1899 in ibid., p. 24; also see Board of Rev. (Land Rev. & Sett.) No. 29, dated 18 May 1927; A. Vipan, *Scheme of Road Development for the Madras Presidency* (Madras, 1935).
35. Letter R. Dis. No. 326 dt. 21 Dec. 1899 from W.O. Horne, Collector & Pres. Dt. Bd., Vizag; Letter dt. 2 Dec. 1899 from H.D. Taylor, Pres. Dt. Bd., Ganjam and Letter Dis. No. 235, dt. 21 Aug. 1900, Gen. (VCR), pp. 41-2 and 14-16; for more information also see *Vizagapatam District Gazetteer* (Madras, 1907); D.F. Carmichael, *Manual of the Vizagapatam District* (Madras, 1868).
36. A. Vipan, *Scheme of Road Development*, pp. 7-8 and 94-6; Board of Rev. (Law Rev. & Sett.) No. 28, dt. 18 May 1927; GOM., G.O. No. 548, Development Dept. dt. 4 April 1932; W. Francis, *Vizagapatam District Gazetteer* (Madras, 1907).
37. See *Review and Returns of See-Borne Trade and Navigation of the Madras Presidency, for years 1899-1900, 1903-04* and *1915 to 1920*; GOM, G.O. No. 1125 Rev. dt. 5 Nov. 1900; G.O. No. 2385 Rev. dt. 22 June 1918; G.O. No. 560; Development Dept., dt. 26 April 1934; G.O. No. 1982, Rev. dt. 24 Sept. 1900; G.O. No. 882, Rev. dt. 11 Sept. 1906; G.O. Nos. 11 and 267 Rev. dt. 9 Jan. and 22 March 1900; G.O. No. 158 (Rev. Sep.) dt. 9 Aug. 1918.
38. From the Honourable D. Arbuthnott, Collector of Madura, to the Acting Secretary to the Board of Revenue, dated Pereacolum, 27 April 1871, No. 155, in Proceedings of the Board of Revenue, No. 3284, 5 Aug. 1871, pp. 5763-4.

39. Observation of T.A.N. Chase, Collector of Kurnool sent to the Acting Secretary to the Board of Revenue, dated 27 May 1871, No. 177, pp. 5751-5, in ibid.
40. Ibid.
41. G.N. Rao, 'Changing Conditions and Growth of Agricultural Economy in the Krishna and Godavari Districts, 1900-1939' (unpublished Ph.D thesis, Andhra University, Waltair, 1973), p. 16.
42. Letter from H. Morris, the Collector of Godavari district to the Board of Revenue, No. 314, dated 28 Sept. 1863, PBOR, No. 7570, dated 1 Dec. 1863, p. 6992.
43. Henry Morris, *Manual of the Godavary District*, Madras, 1878, p. 85.
44. Letter from H. Morris, Collector of Godavari District to the Board of Revenue, No. 314, 28 Sept. 1863, PBOR, No. 7570, dated 1 Dec. 1863, p. 6991.
45. B.H. Baden-Powell, *Memorandum On Forest Settlements In India* (Calcutta, 1892).
46. Report of the Forest Committee, dated 15 May 1882, in G.O. No. 1259, 16 Nov. 1882, p. 2, taken from Selections from the Records of the Madras Government, New (Revenue) Series, No. X, A Collection of Papers Relating to the Administration of the Forest Department (Government Press, Madras, 1915).
47. Public Petitions to the Forest Department, Madras Presidency, 1862, 1865, 1868, 1873, 1876, 1878, 1879, 1885, 1887, 1888 and 1890.
48. R. Dis. 19/17, signed 24 Nov. 1918, Visakhapatnam District Collectorate Records.
49. R. Dis. 38/19, ibid.
50. R. Dis. 41/17, ibid.
51. R. Dis. 51/17; R. Dis. 28/18 and R. Dis. 12/17 in ibid.

Three Issues from a CPR Management: Village Forestry in Post-colonial South Asia

Arun Bandopadhyay

In a recently published monograph entitled *Life on the Edge: Sustaining Agriculture and Community Resources in Fragile Environments* (OUP, Delhi, 2001), N.S. Jodha has focused on a variety of viewpoints for the study of Common Property Resources (CPRs) in post-colonial South Asia. Such a wide variety partly stems from the broader definition of the CPRs that Jodha makes. In this definition, CPRs are of course taken as those non-exclusive resources in which groups of people have coequal rights, but these rights are not seen in terms of legal ownership only. Consequently, even where legal ownership rests with some other agencies like the Revenue Department or the Forest Department (in the case of wastelands or protected forests, for example), *de facto* CPRs belong to the village community or different groups of the forest communities.[1] Viewed from the broader definition, CPRs in Indian villages include community pastures, different kinds of forests including community forests, wastelands, watershed drainages, village ponds, rivers and rivulets with their banks and beds, both from a *de jure* and a *de facto* rights framework.

The disregard of CPRs in rural development programmes as well in effective forest management in post-colonial South Asia is most marked. In terms of the decline in area, the depletion of the CPRs in India from 1951 to 1981 has been measured to vary from 31 to 55 per cent in different regions. In one of our study areas (Tamilnadu), the decline is stated to be 50 per cent.[2] Three factors are mentioned as causes. First, gradual extensions of private field borders took place through outright grabbing of CPR plots individually or collectively.

Second, government also made distribution of CPR land as private land in the name of land reform in the last fifty years. Third, CPR area was also curtailed by various agencies of government like the forest departments or the panchayats in the name of development or resource harnessing.

Recent researches have mostly focused on the management and productivity of CPRs in post-colonial India and in the process are involved with issues of environment and ecology, poverty and equity, institutional and technological interventions, rural factionalism and participatory development, grass roots democracy and bureaucratic perceptions, etc. In order to have clearer, more focused view of the process of change in this area, this paper concentrates on the history of 'village forestry' in post-colonial South Asia, with special reference to Tamilnadu and West Bengal, from the triple perspectives broadly grouped as 'ecological', 'technological' and 'participatory'. The argument is not only that there was a complex interplay of these issues but that they also overlapped and at the same time were interdependent. A historical understanding of the subject with reference to development in the last fifty years is therefore in order.

I

From the middle of the nineteenth century, 'scientific forestry' developed in India under the colonial auspices, which could lead to the artificial demarcation of 'reserved', 'protected' and 'village' or 'community' forests as they were named in various parts of India.[3] However, these discourses on scientific forestry could neither stop the progressive commercialization of forestry of various forms and the marginalization of its resident communities. Indeed, the administrative infrastructure that was developed during the colonial rule proved grossly inadequate in the post-independence era, both from institutional and ideological perspective.[4] There was a renewed emphasis on social and economic development in an altogether different plane in the context of a rapidly growing population. There was a concern for deforestation and development, as exemplified from the official emphasis on 'Vana Mahotsava' and 'Grow More Food' in wastelands in the early 1950s. As early as 1948, the Government of India (hereafter GOI) urged for tree planting as a national festival or Vana Mahotsava in different states in the month of July and the Assistant

Secretary, Ministry of Agriculture, GOI wrote to the Secretary, Development Department, Madras on 23 July 1948:

> I am directed to forward herewith the copy of appeal issued by the Honourable Minister for Food and Agriculture. . . for plantation of quick-growing trees in villages on the 15th August 1948, the first anniversary of Independence Day. As the plantation of quick-growing trees will result in making available to the cultivators a suitable fuel in place of cow-dung now burnt for fuel and thus help to release the latter for its utilization as manure for increasing agricultural production, it is requested that the provincial and state governments may give wide publicity to the above appeal and render all such assistance as may be necessary to the villagers in these areas for observing tree planting programme on the 15th August 1948.[5]

Two years later, the Madras government argued that if the wanton destructions of the vegetable cover of the soil was to be avoided, 'it (wa)s necessary to arouse consciousness among the masses regarding the value of trees and instil in them a zeal for planting more trees and protecting them and an adoration for these gifts of Nature which enrich the earth'.[6]

However, the Vana Mahotsava was not successful, but continued as a ritual for about ten years. The important fact was that it failed not only 'ecologically' or 'technologically' but people's participation was 'indifferent'. In the 1960s and 1970s, farm forestry of various types was introduced in different states of India for the partial greening of the countryside. Here the farmers mainly carried out block planting of their wastelands but over-dependence on eucalyptus was ecologically degrading and the lack of dissemination of market intelligence in terms of prices and products was economically discouraging. The group farm forestry and the joint industrial forestry also met the similar fate. In the latter case, many private companies went into liquidation, as the joint project, based again on eucalyptus and bamboo, was not commercially successful. During the 1970s and 1980s, forest departments of various states were preoccupied with large, plantation-oriented social forestry projects, backed by central government, foreign donor agencies and various NGOs. For example, Bihar state government, backed by the funding of Swedish International Development Authority (SIDA), encouraged the programme for the establishment of nurseries and fast-growing tree plantation in the 1980s, keeping however the indigenous community separate and isolated from forest management initiative.[7] These limitations of social forestry project became apparent over time, encouraging some planners to devise alternative methods of management. Joint Forest Management (JFM), devised as Forest Protection Committees (FPCs) as early as in the 1970s in certain parts of West Bengal, only became widely known in the 1980s.[8] 'In 1988 and 1989, Orissa and West Bengal passed state resolutions recognizing the validity of community forest protection. In June 1990, the Government of India passed guidelines notifying the exclusive rights to forest products be extended to those villages effectively protecting public forest lands. By 1994, sixteen states passed similar orders.'[9]

Thus, the history of village forestry in India, from Vana Mahotsava through farm forestry and social forestry down to the joint forest management, makes an interesting reading of social changes in well over fifty years. It is interesting not only because of the ecological and technological aspects of change, but also because of participatory and by that matter 'good governance' aspects of it. Since our case studies are taken from Tamilnadu and West Bengal, it is important to begin with a general description of forest management after independence in these two states respectively.

There were various working plans to stop forest felling even during the colonial period. After the Second World War, development schemes were initiated in most of the states in India in addition to regular working plan prescriptions. In the post-independence period, these arrangements were thought to be on a meagre scale compared to the needs of the situation. There were large-scale fellings in zamindari and other private forests and even landowners began felling their trees under the inducement of high prices obtainable for wood. Thus, considerable areas of forests of former princely states, of former zamindaris and even of private ownership came to stand in urgent need of special measure of large-scale rehabilitation. The extension of cultivation encouraged by government-sponsored Grow More Food campaigns also resulted in much clearing of forest areas. Inevitably, the pressure of a rapidly increasing population, the improvement demand of the forest for wood and the growing imbalance of the land under permanent vegetation (forests) and land under cultivation highlighted the crucial issue of forest management in ecological and economic terms in both the states of Tamilnadu and West Bengal.

In Tamilnadu, development work other than normally prescribed in the department's working plans began with the First Five Year Plan.

The panchayat forests and some zamindari forests were taken over. Certain soil conservation experiments were made in the Nilgiris. Casuarinas, dry fuel spices, soft woods, wattle and blue gums were planted. Measures of forest conservancy and related prescriptions prohibiting goats in forests, limiting grazing incidence to the capacity of fodder resources, regulating supplies of small timber, fuel and other produce to the villagers, etc., were introduced. Documents suggest that the government was anxious to remove in some cases serious grievances caused by forest conservancy, as a sort of liaison between forest and people but its limitation was obvious. Moreover, expenditure on forestry and forest administration was confined to a small part of the revenue derived from the forests.[10] Indeed, the success of the department was mainly measured in terms of its annual surplus only. The result was that forests especially in dry and naturally poor localities in Tamilnadu suffered progressively for want of special works of 'improvement' meaningful to the people.

A way out was sought by the schemes under the Second Five Year Plan. Forestry schemes such as commercial and industrial plantation, rehabilitation of degraded forests, extension of forestry by way of tree planting along roads, canal banks, rail roads, river margins, preservation of wildlife and better housing and other amenities of staff, labour and forest tribes were made in certain measure. However, forest departments were mainly concerned with afforestation of eroded and denuded hills and forest working plans were mainly providing for the formation of timber population and even that on a limited scale necessary for ensuring regeneration.

During the Third Five Year Plan, work was expanded systematically to areas outside the regular forests, by way of extension forestry, involving extensive planting of trees, in grove as well as in wastelands. An important objective was the establishment of farm and community forestry that began to be explored during the 1960s and early 1970s. The most significant among the early initiatives in this venture was found in Tamilnadu.[11] Farm forestry scheme were extended to tree planting and forestry activities to areas outside departmental forests. The foremost plan of the project was to create local fuel resources and thus free the agricultural population from dependence on village forests for small requirements. Farm forestry subsequently developed into social forestry that meant the management and protection of forests and afforestation of barren lands with the avowed objective of environmental, social and rural development. Its aim was to create goods and services for the consumption, primarily of rural poor. Here the forest department was expected to play the role of educationist and catalyst by providing the needful technical and other services. In Tamilnadu, farm land below a tank and/or watered by a system of canals are known as *ayacut*, whereas the village common land for grazing is known as *poramboke* lands. The main components of social forestry in Tamilnadu were plantations on tank foreshores and *poramboke* lands. Here the central concept that evolved over time was to create forest-based resources in rural areas through the overall improvement of local people in their capacity as individuals as well as members of the village community.

The experiences of West Bengal in forest management in the early years after independence were similar to those of Tamilnadu. One outstanding feature was the taking over possession of the most of private forests of the state by the Forest Department consequent on the implementation of the West Bengal Estate Requisition Act, 1953 (Act I of 1954) and also the transfer of forest areas of Manbhum and Purnea from Bihar to West Bengal, with effect from 1 November 1956. Here also a great drive was made to Vana Mahotsava and wildlife preservation activities during 1955-60.[12] The gross area under forest was 13.3 per cent of the total geographical area of the state. In 1960, the total area of forest of all classes in West Bengal stood at 4,535 sq. miles of which the reserved forests, protected forests and other forests comprised 2,691 sq. miles, 1,573 sq. miles and 271 sq. miles respectively. The tempo of development work in forest was reflected in the increased supply of timber to the Indian Railways and to various government departments and private wood-based industries like plywood, match wood, packing case, etc.

One result of this policy of use of forest was the progressive degradation of it. In 1981, West Bengal had the highest population density of any state in India and lowest per capita forest area (0.02 ha. per person). Rural need for fuel wood and non-wood forest products resulted in uncontrolled exploitation that threatened the very existence of the remaining forests of the state.[13] Conflicts and mistrust between foresters and rural communities increased. However, a novel solution developed as a result of the unique activities of a silviculturist (an Indian Forest Service officer) in 1972 in eleven villages surrounding Arabari, 30 km. to the north of Midnapore, followed by the development of the first Forest Protection Committee (FPC).[14] Here the villagers agreed to protect the natural *sal* forest from fuel woodcutting in return of 25 per cent of all revenue generated from the sale of firewood and timber. The idea was reshaped in mid-1980s to cover the degraded forest areas of Bengal in participatory management, which is known as Joint Forest Management (JFM).[15] In JFM, political support has been enlisted in an institutionalized form by often tying it up with the panchayats. It has been estimated in 1997 that there were 3,112 FPCs under JFM in West Bengal, holding over an area of 5,21,118 ha. with 3,74,322 beneficiary families.[16] Thus, in West Bengal foresters were made accountable to community groups, micro plans were introduced, and a new analytic and administrative framework for supporting those plans were also created increasingly since the middle of the 1980s.

II

It is now important to examine the detailed schemes of Social Forestry and Joint Forest Management from the respective experience of Tamilnadu and West Bengal. For the sake of clarity and better understanding, we are concentrating on specific projects. For example, as far as Tamilnadu is concerned, we are focusing on SIDA-assisted Social Forestry Project, Phase II (1988-9 to 1992-3). In the case of West Bengal, however, we are paying attention to a few areas with a few projects only. While the re-edited apprised document is our main source for Tamilnadu,[17] several detailed studies of JFM are our main sources for West Bengal.[18]

The second phase of Tamilnadu Social Forestry Project (SFP) came as a joint endeavour of the representatives of the Government of Tamilnadu, Government of India and the SIDA Appraisal Mission. The project was originally launched with SIDA assistance in 1981, and irrespective of the fact that community involvement fell short of the expectations in SFP I, it continued to form an integral part of SFP II. It was also decided to widen the base of social forestry to meet the rural forest-based product needs. There were five main project components, i.e. (a) *Tree Patta Scheme* for landless women in panchayat land, (b) *Agro-forestry in Dry Land* for resource-poor farmers, (c) *Community Wasteland Development Programme* with multi-purpose plantations on government and panchayat wastelands, (d) *Interface*

Forestry Programme in degraded reserved forests with improvement of adjoining revenue, *poramboke* and private lands, and (e) *Seedling Production and Extension Programme* through *kissan* nurseries, appropriate extension services, application of research and development work, etc.

It has been estimated that block plantations established on community lands in Tamilnadu from 1960 to 1981 had cumulative area coverage of 1,33,033 ha. At this juncture, the SFP I had the following main objectives, i.e. to increase supplies of fuel wood in rural areas and to provide poles, bamboo, small timber, fodder, grass, fruit, oil seeds and other minor forest products, to establish plantations to satisfy on a long term basis the wood requirements of the villagers, to gradually shift responsibility for protection and harvesting of woodlots raised outside forest lands, from Tamilnadu Forest Department to village panchayats, and to help economically weaker sections to raise plantations in their landholdings unsuitable for agriculture.

However, the defects of SFP I became apparent when project evaluations were made. First, extensive areas of plantations were established on tank foreshores, but much less on *poramboke* hills and plains. Second, a large number of seedlings were distributed for private planting but the greater proportions was planted by bigger farmers and landowners. Finally, effective participation of the community on social forestry did not come up to the expectation because of the conflict of interest within the villages and because the villagers found the administrative procedures, by which plantations were handed over, complex and unacceptable.[19]

In SFP II, several corrective measures were taken. Improved quality and increase in productivity was one such measure. Special endeavours were made to motivate and assist the communities to take over the management of plantation assets created. Stress on fodder production and plantation of fruit-bearing trees were another emphasis. SFP II also aimed at implementing social forestry in reserved forests. Two of the components of the project, i.e. Tree Patta Scheme (TPS) and Interface Forestry (IF) were completely new while others like Agro-forestry in Dry Lands and Community Wasteland Development Programme that had been in existence now contained special features and new organizational arrangements to cope the changes.[20]

The TPS, for example, was evolved to make social forestry a

people's movement and to ensure better survival of seedlings planted in communal *poramboke* lands in the blocks in Tamilnadu. Tree *pattas* were given to local, landless, rural poor women following IRDP norms of 'poorest among the poor', the Scheduled Castes and Scheduled Tribes families forming not less than 50 per cent of the total beneficiary women. 'Each beneficiary was given the right to trees planted in one acre (in case of block plantations or one kilometre (in case of strip plantations) in Panchayat land.'[21] As regards Interface Forest (IF) in Tamilnadu, it was that part of degraded reserve forest, which formed the direct and indirect benefit flows to rural communities in its command area. The Interface Forestry Programme (IFP) aimed at treating and managing such a catchment to ensure increased benefits to improve the living conditions of the people in adjacent rural areas. The IFP would 'necessarily have to be multi-disciplinary and comprehensive in order to meet the multiple objectives of forest cover, water management and meeting the labour and forest product needs of the associated communities through plantations and soil conservation work'.[22] The concept of IF went 'beyond reforestation and water management' to meet the labour and forest product requirement of the people: it sought 'their active participation in the planning, implementation and maintenance of the programme'.[23] Third, the new component in agro-forestry in dry lands was the approach made towards small and marginal farmers in different agro-climatic zones with low rainfall, 'to provide them free seedlings up to a total of 500 each and of species according to farmers' preferences'.[24] Finally, while the main objective of the community wasteland and development programme was to establish self-sustaining assets on community lands including PWD tank-beds and tank foreshores for the benefit of the villagers, the main strategy was to improve productivity of plantations (rotational and non-rotational) and to introduce management systems that could be maintained on a sustained basis by local bodies.[25]

In West Bengal, a number of field research findings highlight certain crucial aspects of JFM. Indian Institute of Bio-social Research and Development (IBRAD) and the Department of Forest, Government of West Bengal in their study of *sal* regeneration in east Midnapore show that there has been remarkable regeneration in the forest lands in the area after the formation of FPCs under JFM programme.[26] It is important to note that increase of vegetation cover has enhanced Non-Timber Forest Products (NTFP) production, which is important for the sustenance for the people as well as the village economy. P. Bhattacharya, S.B. Roy and R.K. Bhadra suggest that very few families within the FPCs have started inter-cropping on the protected forest lands in Baikanthapur Division in north Bengal. 'The community as well as the Forest Department is not very clear about the related factors like suitable soil conditions, supply of seed in proper time, manure, insecticide, proper spacing method, suitability of the crops for the area and season, aftercare and protection from wild animals, etc. Few families have got better returns and some of them have even faced losses.'[27] K.C. Malhotra and others made a study on the role of NTFP in the village economy of south-west Bengal. A total of 216 households belonging to tribal and caste population from 12 FPCs in Jamboni range of Midnapore district were studied. The local communities used the forest products for various purposes. This use of NTFP from the forest was mainly concentrated on 73 plants including trees, shrubs, climbers, fungi, grass and two animal species. The 'maximum utilisation of NTFP was found to be used as food, followed by fuel, medicine, etc. Although most of the species were commonly used by tribals and the caste population, some of the species like *mahua* flower, *bakhar* root and the *karkut* (ants) were consumed exclusively by the tribals.'[28] As regards the food items, *kendu* fruits were mostly raised, followed by mushroom, tubers and leafy vegetables. However, there was a genuine problem of marketing infrastructure for the disposal of the product, as exemplified by a case study of mushroom.[29]

The role of women in utilization of natural resources has been highlighted by a study of Mitali Chatterjee.[30] It is argued that the government order in 1989 on FPCs in West Bengal did not recognize the role of women in forest protection, a mistake that was rectified by a G.O. dated November 1991. As primary and more frequent users of forest resources, women played an important role in the management and decision-making and were formally recognized as voting members of local management groups. But there was a paradox in women's involvement in forest resources, which could not be resolved without their further engagement. As traditional users, they could readily identify offenders, but by an extension of the argument sometimes they could become the offenders themselves. Chatterjee writes,

This is significant in light of the fact sometimes critical situations develop wherein male members of the forest protection committee are unable to take any action against the female offender for fear of consequences. Here an example deserves to be added. Dukhu Tudu, an old tribal man of Dahi village under the Khajra forest beat of Midnapore district of West Bengal, once came upon a few women who were involved in indiscriminate cutting of trees in nearby forest. Dukhu tried to prevent them from damaging the forest but women refused to listen to him. He tried to stop them by snatching away all their axes, but in turn, it was he who was blamed for misconduct. Those women complained to their husbands against Dukhu Tudu, accusing him physical attack and molestation. The men from the village of these women offenders then came in a group seeking revenge with the result that Dukhu Tudu suffered serious head injuries. This is no isolated case; there have been many such incidents and consequently men have not dared to catch women offenders. Even some of the staff members of the forest department are scared of preventing women from such unlawful acts. But if women themselves are involved and given the responsibility of forest management, they can work very effectively, and the problem of forest protection and economic status of women can be tackled simultaneously.[31]

The combination of male and female participation in certain types of work is also noticeable. Madhumati Dutta and Manasi Adhikari's study of *sal* leaf plate making in Sabalmara, west Midnapore shows that it is a household industry, largely done by womenfolk, although male members of the family also participate. However, the main problem faced by the producers is that they receive a very low price for the product and the return to labour is the lowest in the industry.[32]

Local institutions also do count in effective management of forest resources. But what was their exact role and what did the people expect from them in West Bengal in the 1990s? S.K. Maity and K. Bhattacharya have enquired the institutional role of the panchayat in JFM programme into some villages of Khargpur and Jhargram Social Forestry Ranges in Midnapore.[33] It has been found that about 58.8 per cent of the respondents observed that the panchayats played no role in farm forestry programme, while 24 per cent admitted that they encouraged farmers through meetings and group discussion to adopt the programme and another 22 per cent perceived that panchayat distributed seedlings, fertilizers and insecticide supplied by the Forest Department. However, a more active involvement of the panchayats in JFM was expected by a predominant number of respondents (81.6 per cent). The study of Mridul Manna and a team of researchers from Regional Research and Study Centre (RRSC), Midnapore, shows that JFM was encouraging in a number of villages in the vicinity of forests in the district of Bankura, West Midnapore and Purulia not only because of the 'greening experience' but also of their 'participatory experience' in this programme.[34]

III

The chequered history of village forestry in post-colonial South Asia—its shifting emphasis, changed directions and continuing experimentation—is interesting and important as it centres on certain cardinal issues of contemporary social change. The three issues identified in this paper are grouped in the ecological, technological and participatory concerns of the localities and communities but they also contain within their respective folds subtle issues of poverty and equity, institutional and technological interventions and grass-roots democracy and bureaucratic perceptions.

The ecological imperatives of the village forestry in its various forms stem from the experience of diversified resource use of land and forest products under highly variable climatic conditions. There was a time-tested requirement of sub-marginal or fragile lands under low intensity uses like natural vegetation as against annual cropping. Consider the case of Interface Forestry or that of Agro-forestry in Dry Land or the Community Development Wasteland Programme in Tamilnadu: in each case marginal lands, wastelands or lands in degraded forests were involved for 'improvement' for the 'good' of the community However, a distinction has to be made between plantations of commercial timber/wood for use of government, contractors or people on the one hand and the more diversified use of forest products for food and non-food purposes. For example, the enthusiasm 'to plant eucalyptus declined after 1986' as they failed to generate profitable returns and proved to be less sustainable.[35]

The technological focus was addressed to break a vicious circle involving degradation, neglect and further degradation of the broadly defined CPRs. It is to be seen that productivity was not the only question but there were other related issues. Persistent gaps between the perspectives of the technologists and resource users provided an issue to be constantly handled by the resource managers. Other difficulties arose from frequent high priorities to commercial considerations by the technologists ignoring the CPR perspective. The emphasis on NTFP was one such perspective that was duly incorporated in JFM programme in West Bengal.

The participatory question was very important as it highlighted the grass-roots level democratization of resource management system and participatory development process. The question, however, had its bearing on the ecological and technological aspects of change in forest management in contemporary South Asia. For example, much useful ecological knowledge and the idea of appropriate technology came from indigenous communities engaged in forest management.[36] Participatory perspectives also reduced the cost of policing and subsidizing the resources necessary for other forms of forest management. But there must be incentives.[37] Though local responsibilities very often rest with the poor in the management of local resources, participatory development can only grow with institutional assistance, within a favourable political space.[38] However, most importantly, the participatory perspectives highlight two cardinal aspects of social changes in contemporary India. First, as an offshoot of grass-roots democratization involved in the process, other issues like gender or equity became more and more prominent. Second, it also envisaged a significant change in the attitude of forester involved in the formation of participatory forest management. For instance, only very recently it has been appreciated that generally 'both dependence and influence of forest guards on villagers as well as the reverse situation has a critical influence on the implementation of any forest policy'.[39] Indeed, much of the explanation of the gap between policy and implementation in forest management in contemporary India may be found in the hitherto unexplored interactions, conflicts and negotiations that occurred between the ground-level implementer and forest society at large. Naturally, the expectation from a forester phenomenally rose during this period. It is no wonder, therefore, that the Training Manual of Joint Forest Management prepared by S.B. Roy and Mitali Chatterjee on behalf of IBRAD, New Delhi, retain a full chapter each on their 'Social Skills' and on their 'Motivation'.[40] These changes are unusual and they are meaningful only with the reference to broader process of social change in post-colonial South Asia.

NOTES

1. See N.S. Jodha, *Life on the Edge: Sustaining Agriculture and Common Property Resources in Fragile Environments* (Delhi, 2001), p. 120.
2. Ibid., p. 135.
3. See Madhav Gadgil and Ramachandra Guha, *This Fissured Land: An Ecological History of India* (Delhi, 1992), pp. 113-45; and *Ecology and Equity: The Use and Abuse of Nature in Contemporary India* (Delhi, 1995), pp. 148-75.
4. See Subhabrata Palit, 'Indian Forest Departments in Transition', in Mark Poffenberger and Betsey McGean (eds.), *Village Voices, Forest Choices: Joint Forest Management in India* (Delhi, 1998), pp. 210-29.
5. See Development Department, Madras, G.O. 4123, dt. 6.8.1948, Tamilnadu Archives (hereafter TNA).
6. See Food and Agriculture Department G.O., MS No. 1108, dt. 7.6.1950, TNA.
7. See Mark Poffenberger, Betsy McGean, Arvind Khare, 'Communities Sustaining India's Forests in the Twenty-first Century', in Poffenberger and McGean (eds.), op. cit., p. 32.
8. See Mark Poffenberger, 'The Resurgence of Community Forest Management in the Jungle Mahals of West Bengal', in David Arnold and Ramachandra Guha, *Nature, Culture, Imperialism: Essays on the Environmental History of South Asia* (Delhi, 1995), pp. 350-4.
9. See Poffenberger, McGean, Khare in Poffenberger and McGean, op. cit., pp. 26-7. These guidelines were transformed into legislation, which was given wide publicity and possible extension in meaning in the 1990s. As Justice P.N. Bhagwati (former Chief Justice of India) wrote in April 1996, 'I find that, unfortunately, laws for the protection of the forests are highly theoretical and unjust. . . . These laws are conceived in the offices of the bureaucrats and the chambers of the concerned minister, regardless of the situation prevailing at the ground level and totally ignoring the lifestyle of the forest denizens. The answer lies in including the people residing in the forest regions and eking their daily living from forest produce, taking them into confidence and giving them an effective voice in the policy decisions affecting forest management.' See Justice Bhagwati's Foreword to the study sponsored by Rural Litigation and Entitlement Kendra, Dehra Dun, *Community Forest Management in Protected Areas—Van Gujjars, Proposal for the Rajaji Area* (Delhi, 1996), p. i.
10. See D. Janaki, 'History of Forest Administration in Tamilnadu, 1935-1967' (unpublished Ph.D thesis), Department of Indian History, University of Madras, 1990, p. 144.
11. See G. Foley and G. Barnard, *Farm and Forestry* (London, 1984), pp. 14-15.
12. See Forest Department, Government of West Bengal, *Review of the Forest Administration in West Bengal from April 1, 1955 to March 31, 1960* (Alipore, 1964).
13. See Mark Poffenberger, 'Joint Management for Forest Lands: Experiences from South Asia, a Ford Foundation Program Statement' (mimeo, Delhi, January 1990), p. 9.
14. Ibid., p. 10. See also Mark Poffenberger, 'The Resurgence . . .', in Arnold and Guha, op. cit., pp. 351-2.
15. 'In the early 1980s, recognizing the success of Arabari and a few other villages

where management agreements with forest communities were being made, some senior forest officers began to encourage field staff to pursue similar negotiations in wider areas throughout the south-western part of the state.' See Poffenberger in Arnold and Guha, op. cit., p. 353.

16. See S. Palit, 'Joint Forest Management in West Bengal', in *Annual Report 1996-97 and Proceedings of Annual Workshop, Regional Centre, National Afforestation and Eco-Development Board* (Ministry of Environment and Forests, GOI, 1997), p. 67. Hereafter, the volume is referred to as *Annual Report 1996-97 and Proceedings.*
17. See 'Social Forestry Project Tamilnadu, Phase II (1988/89-1992/93), Re-edited Appraised Project Document' (Madras, mimeo), TNA. This document is hereafter referred to as SFP II.
18. These sources are cited below, separately. For a general review of areas other than West Bengal, see Nandini Sundar, Roger Jeffery, and Neil Thin, *Branching Out, Joint Forest Management in India* (Delhi, 2001).
19. See SFP II, op. cit., p. 10.
20. Ibid., p. 16.
21. Ibid., p. 18.
22. Ibid., p. 34.
23. Ibid.
24. Ibid., p. 23.
25. Ibid., p. 28.
26. See IBRAD and Department of Forest, Government of West Bengal, 'Study of the Vegetation Dynamics in Regenerating Sal Forest of Moupal Beat, East Midnapore Forest Division, District Midnapore, West Bengal', in S.B. Roy and Mitali Chatterjee, *Joint Forest Management, A Training Manual* (Delhi, 1994), pp. 145-6.
27. See P. Bhattacharya, S.B. Roy and R.K. Bhadra, 'Scope of Intercropping Practices on Protected Forest Lands of North Bengal—Ecological and Economic Perspective', in ibid., p. 148.
28. See K.C. Malhotra, Debal Deb, M. Dutta, T.S. Vasulu, G. Yadav and M. Adhikari, 'Role of Non-Timber Forest Produce in Village Economy', in ibid., p. 150.
29. '. . . in the marketing arena, there is a need for marketing assistance programmes, which would help in giving information regarding local and domestic markets and new opportunities, export assistance network, and trade fairs/exhibition'. See Vanita Suneja, S.B. Roy, and J.K. Das, 'Marketing of Mushroom', in S.B. Roy and Mitali Chatterjee, op. cit., p. 153.
30. See Mitali Chatterjee, 'Women in Joint Forest Management: A Case Study from West Bengal', in Mitali Chatterjee and S.B. Roy (eds.), *Reflections from Training on Gender Issues* (Delhi, 1994), pp. 47-68.
31. Ibid., p. 57.
32. See Madhumati Dutta and Manasi Adhikari, 'Sal Leaf Plate Making in West Bengal: A Case Study of the Cottage Industry in Sabalmara, West Midnapore', in Roy and Chatterjee, *Joint Forest Management*, op. cit., p. 149.
33. See S.K. Maity and K. Bhattacharya, 'Farm Forestry', in *Annual Report 1996-97 and Proceedings*, op. cit., pp. 71-2.
34. See Mridul Manna, 'Some Experiences of Afforestation involving People in Bihar and West Bengal', in ibid., pp. 75-6: 'A time has come when RRSC workers need not go to the village to motivate villagers. Rather the villagers come to RRSC workers to motivate them so that they could undertake new ventures according to

villagers' need and spread to areas where there is lack of afforestation.'

35. See N.C. Saxena, 'Farm Forestry in different Agro-Ecological Regions of India', p. 4, cited in N.H. Ravindranath, Madhav Gadgil and Jeff Campbell; 'Ecological Stabilization and Community Needs: Managing India's Forest by Objective', in Poffenberger and McGean (eds.), op. cit., p. 294.
36. Consider, for example, the arguments of Dev Nathan and Govind Kelkar for the establishment of forest communities' property rights over forests in 'Case for Local Forest Management, Environmental Services, Internalisation of Costs and Markets', *Economic and Political Weekly* (hereafter *EPW*), Vol. XXXVI, No. 30 (28 July-3 August 2001). For them 'this local knowledge of the forest communities is indispensable to developing higher productivity-based multiple-use forest systems, based on ecosystem rather than single product management. Community management, or local interest management, rather than management by forest departments, can best harness this knowledge for effective forest management.' Ibid., pp. 2839-40.
37. This seems to be a common lesson from the environmental projects elsewhere. See Michael Stahl, 'Environmental Rehabilitation in the Northern Ethiopian Highlands: Constraints to People's Participation', in Dharam Ghai and Jessica M. Vivian (eds.), *Grassroots Environmental Action, People's Participation in Sustainable Development* (London and New York, 1995). Stahl concludes, 'The incentives must be directed to households and individuals, as well as to common resources such as village woodlots. Villagers do not plant and tend woodlots productively unless there are rules and procedures for how households can benefit from the collective enterprise by securing fuelwood and construction material for their individual needs.' Ibid., p. 298.
38. See Philippe Egger and Jean Majeres, 'Local Resource Management and Development: Strategic Dimensions of People's Participation', in Ghai and Vivian (eds.), op. cit., pp. 304-24.
39. See Sudha Vasan, 'Ethnography of the Forest Guard, Contrasting Discourses, Conflicting Roles and Policy Implementation' , *EPW*, Vol. XXXVII, No. 40 (5-11 October 2002), p. 4132.
40. See Roy and Chatterjee, op. cit., Chapters 5 and 6. Consider, for example, the two observations made in the Manual. 'Even if few persons agree to form FPC it is very difficult to unite. Villagers have different group affiliations, political ethos, cultural norm and are influenced by several other groups. This causes conflict and may be resolved through social skills which have been developed based on several researches on social science. Once the foresters learn to manage man, they can manage the forest flora and fauna more effectively.' Ibid., Chapter 5, 'Social Skills', p. 59. See also Chapter 6, 'Motivation', p. 90: 'To be more precise, people do not destroy forest for want of fuel only. That means people will not stop destroying the forest when one provides them with fuel. . . . One has to examine the whole experiment in a historical perspective where foresters as a government agency have certain norms in their mind that by providing some material or economic gain to the community the problem could be solved. The people called "beneficiaries" have their own perception of government departments and government dole. We may call it the cultural values in this regard. In order to judge the impact of such an event one has to examine the conditions and attitude of the people before and after the programme.'

Contributors

Arun Bandopadhyay is Nurul Hasan Professor of History, University of Calcutta.

B.D. Chattopadhyaya is Professor at the Centre for Historical Studies, Jawaharlal Nehru University, New Delhi.

B.B. Chaudhuri is a well-known historian. He was formerly Professor of Economic and Social History, University of Calcutta.

Amar Farooqui is Professor of History at Hans Raj College, University of Delhi, and formerly a Fellow, Nehru Memorial Museum & Library, New Delhi.

Rajan Gurukkal is Professor and Director, School of Social Sciences, Mahatma Gandhi University, Kottayam.

Atlury Murali teaches at the Department of History, Hyderabad Central University, Hyderabad.

Shishir Kumar Panda is Professor at the Department of History, Berhampur University.

Archana Prasad is at present a Fellow, Nehru Memorial Museum & Library, New Delhi.

P.K. Shukla is a Deputy Director, Indian Council of Historical Research, New Delhi.

K.S. Singh has retired from the Indian Administrative Service. He was formerly Director-General of Anthropological Survey of India.

K.T. Thomas teaches History at Christ College, Trichur.

Kesavan Veluthat is Professor of History at Mangalore University, Mangalagangotri.